URBAN GOVERNANCE AND SMART CITY PLANNING

URBAN GOVERNANCE AND SMART CITY PLANNING

Lessons from Singapore

ZAHEER ALLAM

The Port Louis Development Initiative, Mauritius

emerald PUBLISHING

United Kingdom – North America – Japan – India
Malaysia – China

Emerald Publishing Limited
Howard House, Wagon Lane, Bingley BD16 1WA, UK

First edition 2020

Reprints and permissions service
Contact: permissions@emeraldinsight.com

British Library Cataloguing in Publication Data
A catalogue record for this book is available from the British Library

ISBN: 978-1-83982-107-3 (Print)
ISBN: 978-1-83982-104-2 (Online)
ISBN: 978-1-83982-106-6 (Epub)

CONTENTS

LIST OF FIGURES

ABOUT THE AUTHOR

Zaheer Allam holds a PhD from Curtin University (Australia), an MA (Res) in Political Economy from the University of Sydney (Australia), an MBA from Anglia Ruskin University (UK) and a Bachelor of Applied Science in Architectural Science from Curtin University (Australia). Based in Mauritius, he works as an Urban Strategist for The Port Louis Development Initiative (PLDI) and the Global Creative Leadership Initiative (GCLI) and consults on a number of projects on the thematic of Smart Cities across the African continent and on strategies dwelling in the increasing role of technology in Culture and the Society. Zaheer is also the African representative of the International Society of Biourbanism (ISB), member of the Advisory Circle of the International Federation of Landscape Architects (IFLA) and a member of a number of other international bodies. For his contributions to society, he was elevated, by the President of Mauritius, to the rank of Officer of the Order of the Star and Key of the Indian Ocean (OSK), the highest distinct order of merit in Mauritius. He is also a recipient of a number of other awards and writes extensively on the thematic of sustainability and resilience in cities. He is also the author of *Theology and Urban Sustainability* and *Cities and the Digital Revolution: Aligning Technology and Humanity*.

FOREWORD

LOOK BEYOND THE 'GREEN CITY MIRACLE'

By Professor David S. Jones, FAILA

The urban regenerative process, as advanced in Singapore, is perhaps one of the most interesting precedents internationally, from both academic and professional practice standpoints. It is not simply a tree planning agenda, nor a greening the city agenda. It is more holistic and enlightened about the need to address urban regeneration, landscape healing and climate change/resilience-responsive strategies to the betterment and well-being of both the landscape and Singapore's citizens and visitors.

This strategy addressed all dimensions of liveability including societal, economic and environmental ones well before international recognition and acceptance of sustainable development, the essential message in the United Nations' SDG 11 (Sustainable Communities), was considered and adopted. The Singaporean approach is to use social uplift as a catalyst for urban regeneration. Such is unique in its very endeavour. But one must also acknowledge the governance models that led to this process and the societal transition through the strong leadership of the then Prime Minister Lee Kuan Yew (1923–2015).

In retrospect, Lee has stated:

> Cities cannot just be made up of concrete buildings, tarmac and pavements. It would be depressing and unpleasant to live in. You need to balance that with trees and flowers. This will make Singapore more pleasant to live in.
>
> "Singapore has become much greener, despite increased urbanization. Almost half of Singapore is covered in greenery. We have set aside land for world-class gardens, parks and nature reserves. Many visitors are amazed at our tree-lined roads, and this has become an economic value to us. More importantly, Singaporeans today live in beautifully landscaped housing estates and are able to exercise and enjoy fresh air in the urban oases right at their doorsteps. None of this would have been possible without decades of conscientious planning and commitment (Lee in Soper, 2015).

Lee's strong leadership helped Singapore graduate from a third to a first world country in just under one generation. This endeavour is both remarkable and commendable. While it may be sometimes criticised today by Western democracies and Western authors, one must understand the context of this style of urban governance in respect to both time and place. Post-British and Japanese colonialism saw a Singapore plagued by poverty and poor infrastructures that had scared this society on many levels. The urban governance model applied was led through the willing co-operation of a population whom acknowledged the need for change, transform, prosperity, stability and social cohesion. The interesting aspect of this is how large urban projects of national importance have successfully been devised from inception to delivery, within short time-frames, while engaging in participatory planning. Today, this success is not contested. As a consequence, Singapore enjoys a strong demographic platform that invests significantly in its urban realm and acknowledges the role of technology in furthering participatory planning to ensure a wider social cohesion whereby Singaporeans take ownership of state-led projects. Such is a key in ensuring high liveability levels in urban areas.

While architectural projects can be deployed in a few years, cities, on the other hand, can take generations and longitudinal landscape architecture projects to unfold to unveil their full potential. Though widely acclaimed, if we are to look at Singapore from a timescale, we will realise that the country is still young, is yet to fully establish itself and is now embarking on another exciting transformational journey. As this transformation unfolds, there will be increasing attention from both the academic and professional practice circles on how Singapore's societal and economic dimensions are been crafted and mediated to support, sustain and strengthen human and biodiversity quality of life, which also include the safeguarding of environmental assets.

Zaheer Allam has successfully addressed this topic in this book. Offering an insightful inquiry, he has explained the governance structures and participatory planning key processes as adopted by Singapore and their roles in driving this transformation. Although the same governance structures may be difficult to adopt, in our day and age, in other countries, there are 'lessons to be learnt' from Singapore's transformational journey. Better understanding the past is a key in providing a better outlook on the future, and Singapore may be just innovatively leading the way by offering an inspiration that change is possible – irrespective of the challenges – but also by offering a co-ordinated 'toolkit' of how to effectively enable societal adhesion in urban regenerative processes. Allam has synthesised the Singaporean story carefully, and this book may hopefully act as a guide for others to calibrate the lessons learnt to better adapt to their own cities or countries.

As the world grapples today with challenges of scale, through demographic growth, urbanisation and climate change, we need revised urban governance models. Maybe it is the Singapore model that offers templates and ideas that can inspire us to change ourselves towards alternate urban survival strategies and enable the greater enrichment of humanity.

ABOUT

Professor David Jones has been the Foundation Professor of Landscape Architecture and Planning at Deakin University since 2011. He has degrees in planning, landscape architecture and heritage studies, including a PhD from the University of Pennsylvania. His teaching and research spans across urban planning, landscape architecture, indigenous knowledge systems, regenerative systems and biophilia. His portfolio includes work on the Forest Gallery at Museum Victoria (1995–1996), the Victoria Square-Tarntanyangga Regeneration Project (2017), the *Adelaide Park Lands and Squares Cultural Landscape Assessment Study* (2007) and design scenarios for Gunditjmara lands associated with the Budj Bim National Heritage Landscape and their World Heritage Listed property. He is the co-author of *Geelong's Changing Landscape: Ecology, Development and Conservation* (2019),

Re-casting Terra Nullius Blindness (2017), *Creating Healthy Places: Railway Stations, Biophilic Design and the Melbourne Metro Rail Project* (2017) and *Aboriginal Reconnections* (2013), and has co-contributed significant chapters to the *Routledge Handbook to Landscape and Food* (2018) and the *The Handbook of Contemporary Indigenous Architecture* (2018).

PREFACE

The subject of seeking urban development and economic prosperity in a post-colonial setting is increasingly sought as countries, formerly under the reign of colonial powers, move away from dictatorial regimes to adopt models of democracy. This shift is well received by communities as those witnessed how stronghold regimes can exploit the resources of areas for personal profits, and how this can lead to marginalised communities, hence rendering a fabric of inequity, being a breeding ground for conflicts.

The case of Singapore is interesting in this sense as we see how, immediately after its colonial episode, the country embarked on a transformative journey so successful that the country has managed to move from a third world to a first world country in only one generation. Even if its style of autocratic governance has often been criticised, we need to adopt an objective view and explore how this has led to numerous positive outcomes which have helped the government to provide basic services to an impoverished post-colonial state at an accelerated rate. The success of Singapore is apparent, as today the country enjoys the quality of life – a high-class society.

So how did the country manage this feat? What can we learn from this style of governance? And how sustainable was this transformation? This book looks into those questions and unveils the urban narrative from a historical perspective and explores the key dimensions in the urban policy of the country. While we cannot openly advocate for the establishment of similar governance styles in other countries, the study of the Singapore's success story can help to identify how we could re-calibrate our current governance models in our own context in developments of varying scale and complexity to achieve similar results.

One aspect that has attracted my interest in the formulation of this book is how the local population responded positively to the strong leadership style of Lee Kuan Yew, which begs the question on the validity of complex processes of participatory planning. Indeed, the country would not have achieved this success at this rate if numerous layers of community validation had to be achieved. We then see how prioritisation on certain issues were adopted, where participatory processes were included but only at selected levels, hence achieving community support while allowing for the fast deployment of urban infrastructure and services.

This book thus sets forth to explore the thematic of Urban Governance in the Singaporean context and underlines subjects of interest from the viewpoint of an independent and objective urban practitioner and researcher.

ACKNOWLEDGEMENTS

I am indebted to the Singapore's Centre for Liveable Cities for an invitation for their flagship programme, the 'International Leaders for Urban Governance', in 2018 and later to attend the World Cities Summit in 2019, during which I was introduced to the challenges faced by the country, their incredibly bold solutions and lessons learnt from their transition from a third to first world country. I am also indebted to Sameerah Aumjaud for the editing of this manuscript.

1

THE RISE OF SINGAPORE: AN URBAN PERSPECTIVE

INTRODUCTION

Singapore is a small country, and can be termed as a city, both in terms of size and population. Singapore also carries a rich and vibrant history (Crinson, 2017; Tan, Chuah, & Luu, 2018), which can be paradoxically summed up through despair and massive triumph. Throughout history, Singapore's strategic position on the global map has rendered it a target to the earliest colonial and imperial powers, especially in regard to trade. Singapore eventually became a critical point for the British Empire from which it could access merchant from the larger Asian regions for trade (Findlay & O'Rourke, 2007). Furthermore, the British used Singapore, which was then just a port, as one of its control point for its broader agenda of conquering what Lloyd (1996) calls the Burmese coastline. By capturing such areas, the colony was assured of an expansive trade route covering almost the entire South Eastern Asia. This port as Findlay and O'Rourke (2007) explain was so significant to the British for the aforementioned objectives, such that they were ready to defend and protect it at any cost.

But, as Yong (2019) explains, the history of Singapore and its trading prowess did not start with the landing of the British. Before their arrival, Singapore was already an active trading point and widely engaged with its counterparts in flourishing trade and commerce. However, it was the developments started out by the British that brought about the impetus that marked the beginning of Singapore as a bustling port city, which has maintained its attractiveness as a trading centre to date. But, again, its standing today has little to do with the colonial administrations, but every ounce of success can be credited to the Singaporean and the sound leadership that took over after the departure of the colonialists.

Indeed, by the time the country gained independence, it was so crowded by numerous challenges: massive corruption, pollution, poverty, poor housing, lack of clean water and little infrastructural investments, amongst others (OECD, 2011; Quah, 2017). Unknown to many people, the fortunes of Singapore started to change after Lee Kuan Yew (LKY); the country's first prime minister took over the reins of leadership. From a vast ocean of literature, it is recorded how Prime Minister LKY proposed strong, quantifiable and workable measures that were geared towards uplifting the livelihood and lifestyle of all the inhabitants of Singapore. Today, Singaporeans rightfully take great pride in this uproar. Such pride, as expounded by the Ministry of Foreign Affairs Singapore (2018), emanates from the fact that the country was ranked amongst the most competitive economies globally, due to its very promising and sustainable environment.

Such high standings are based on the practicability of the measures that were agreed upon by LKY, his government and the entire country. And, noting that Singapore is just a small country (approximately 719 km^2) hosting only 5.7 million people, the measures taken were oriented on a city environment. The main reason for viewing the country as a city is that the country is not endowed with any tangible resources, aside from its position as port city (Civil Service College & Centre for Liveable Cities, 2014). However, unlike in most urban cases, Singapore cannot rely on hinterlands of its own to support the challenges it faces. Therefore, Singapore found itself treating neighbouring countries like Malaysia and China as its hinterlands (Tan et al., 2018; Wee, 1995). For instance, Malaysia is the source of Singapore's clean water, though at a price, while China has been very instrumental in providing numerous resources, especially food and building materials as explained by Kolesnikov-Jessop (2010). According to Chye (2014), the measures that LKY initiated revolved around making the country a leader in economic competitiveness, environmental sustainability and liveability status. And, as of the current ranking, those initiatives have been achieved. The Economist Intelligence Unit (2018) ranked Singapore 11th in Asia in terms of liveability.

In regard to economic competitiveness, the country holds the top position, beating economies such as the United States and Hong Kong amongst other global economic giants (IMD, 2019). Pertaining to environmental sustainability, a 2018 report by Environmental Performance Index (EPI) (Environmental Performance Index, 2019) places the country at position 49 out of 180 countries globally. Noting its initial problems with the environment, notably the massive pollution of the Singapore River, the country's ranking is commendable as it holds the leading position in terms of water resource sustainability. From the report, it also ranks at the third position in regard to air quality and takes the first position in respect to household solid fuels use and also the first in terms of household sanitation.

In terms of liveability ranking, basing on factors like infrastructure, education, culture and environment, healthcare and social stability, the country ranks 37 as of 2018, but stands as the most liveable location for East Asian Expats, a position it has held for the past 14 years (Chan, 2018; The Economist Intelligence Unit, 2018). These rankings are credited to the country's effort to improve safety, hence considerably reducing the crime rate (Gallup, 2018; Singapore Police Force, 2016, 2017, 2018). It is also credited for its good and robust infrastructure development as well as its efficient public service. From the very beginning, the Singaporean leadership was based on corruption intolerance, and anyone reported to engage in the vice is punished severely (Yew, 2015b). The public service is also guided by other principles in the likes of pragmatism, commitment to sound institution, community participation and open to working with the markets. Furthermore, the country is guided by Singapore's Liveability Framework (Chye, 2014; Civil Service College & Centre for Liveable Cities, 2014), which outlines the baseline for master planning and the urban governance, which are applied in a strict manner. H. Han (2017) showcases that the master plan encapsulates issues like housing, community development, transportation sector, the economy, recreation and identity that target to enrich the liveability status of the Singaporeans.

SINGAPORE AS THE IDEAL CASE STUDY FOR URBAN REGENERATION

The scale of Singapore brings about a compelling element of interest. First, as noted above, the country is small, and its size is comparative to that of a city. In perspective to scale, according to the United Nations (2016), there are approximately 31 megacities globally that host more than 10 million people and these are by far more populous than Singapore, yet they are not classified as countries. Singapore has an approximate population of 5.7 million people, and as classified by world ranking standards, thus, conventionally, Singapore is only fit to be considered as a large city. Its surface area is another factor that makes it a classical case study to consider. As noted above, by the time it gained independence, only 581 km^2 of its surface area was habitable, but the devotion and passion of the Singaporeans allowed them to reclaim further grounds from the sea to push the habitable grounds to 719 km^2 (Civil Service College & Centre for Liveable Cities, 2014).

Another factor that makes Singapore an interesting case study is its success despite its lack of resource endowment, hence had to intensely rely on its neighbours for almost everything that makes a city liveable. Nevertheless, as noted in a report by the Environmental Performance Index (2019), the country has managed to turn around some statistics in regard to clean water supply. Initially, it was relying almost entire on Malaysian supply for this, but over the years, through an elaborate water cleaning strategy, the country has been able to complement its importation. The EPI report ranks Singapore 13th globally in terms of clean water provision to its citizens. The cleaning process started back in 1977 and entailed the start of the cleaning of the Singapore River, which is now an attractive recreational environment (Tortajada & Joshi, 2014). The country has also managed to ensure food supply sustainability through projects like the Sino-Singapore Jilin Food Zone (SSJFZ), that is, a collaboration between Singapore and China (Ludher, 2016; Tortajada & Hongzhou, 2016).

In terms of social welfare and affordable housing, which is a constant struggle faced by cities globally, Singapore has shown the way. It is reported that Singapore's home ownership stands at over 90%, and this was enabled by the Housing and Development Board (HDB), a government agency that builds compact, mixed-use houses and sells to the locals at subsidised prices (Chia, Li, & Yang, 2017). These houses, unlike in many cities, are compliant with issues like environmental sustainability, especially in regard to energy and water usage and also in use of environmentally friendly construction materials (Kolczak, 2017).

The country's leadership structure and approach are another factor that make Singapore a worthy case study, and this is demonstrated by the number of countries that are borrowing the country's blueprint to duplicate in their own national and urban policies. As noted above, LKY set a pace and standards of leadership and vision that are worth emulating. He guided the country out of corruption, poverty and pollution to achieve the high rankings discussed in the Introduction section. Most cities faced with similar challenges have only struggled, and some have not done much to move out of such quagmires, and by taking the Singapore's challenge, they can ultimately start to experience some differences.

A HISTORICAL PERSPECTIVE

Singapore's history is quite unclear before the third century. Only some tales exist of Singapore during that time, and a verifiable account of the city is only available after the arrival of Raffles, the first British to land on the island. Findlay and O'Rourke (2007) describe this history as obscure and somehow disjointed. Nevertheless, the first account is that of a Chinese record that shows that the current city was just a marshy island at the end of the Malay Peninsula. They had named it *Pu Luo Chung* (a Malay name for an 'island at the end'). The tale twists and Prince Sri Tri Buana (as also known as Sang Nila Utama) of Srivajaya is introduced as the one who discovered the Island in the thirteenth century

(Hack, Margolin, & Delaye, 2010). Leyden and Raffles (1821) explain that the prince was in a hunting expedition when he caught sight of a strange looking animal that he thought was a lion, following which he named the land 'Singhapura', which in Sanskrit dialect means a 'Lion City'. In the fourteenth century, it is said that a number of traders, some from the Mongol Empire and others from China, also described to have made stops in an island which they called *Tamasek* or the *Sea Town*, which is believed to be the modern-day Singapore. During this period, another Srivajaya prince by the name Parameswara is noted to have fled his country to *Tamasek*, and later to Melaka where he gained control and founded the Sultanate of Malacca (Malacca Strait). The sultanate became a successful one, especially in regard to trade, and was a solid trade partner with Singapore, in its previous form. Yong (2019) explains that, indeed, by then, Singapore was already engaged in an intricate web of trade with other autonomous ports and was also a centre for power and politics.

It was the arrival of Sir Stamford Raffles in the nineteenth century that led to the international focus on Singapore. This propelled the island as the regional trade hub as well as a political and power point. Cangi (1993) credits Sir Raffles with, amongst other things, bringing civilisation to the island city, by proposing an elaborate and robust city plan that changed the port from a mere sea city to strategic port city that served as the doorway between the entire Southeast Asia and Britain. Astonished by the political regimes that reigned in the region, and specifically the city, Raffles made it his duty to persuade the locals to adopt such proactive, good practices like accepting democracy, allowing free trade with the neighbours and preaching the need for administration of justice. His motivation was driven by the tyrannical and capricious nature of the rulers ruling the Straits, especially in regard to how they treated their subjects. In respect to his view on trade practices in the area, Raffles was baffled by and disliked the Mercantilism approach by which those who controlled the port employed. He therefore advocated for a free trade system that eventually led to the growth of Singapore as an admirable port city (Donnithome, 2003). Changes in the cities fortunes directly affected its demographic distribution. This led an amalgamation of cultures including Europeans, Chinese, Malays, Arabs, Indians and Bugis tribesmen (Buckley, 1984).

By 1824, five years after Raffles first landed in Singapore, the trading was acquired by the British East India Company, as annual payment by the Sultan who controlled the Islands. By then, the city had an approximate 10,683 inhabitants (Buckley, 1984), but the population was growing fast. By 1826, Singapore joined Melaka and Penang to form what was known as Straits Settlements, but the union did not last long. Due to issues with politics, Singapore separated from the settlement, and in 1867, it became a Crown colony, meaning that it was directly under the British government and not the British East India Company that controlled the Straits Settlements. Turnbull (2009) explains how political, economic and social issues like piracy, unfair taxation, unstandardised currencies and inhumane transportation of convicts who also experienced delayed judicial process are amongst issues that prompted Singapore to be separated from the Straits Settlements. From there, in 1867 (BBC News, 2018), it became a crown colony; meaning that it was directly under the British government and not the East India Company that controlled the Straits Settlements. Buckley (1984) shares that merchants, especially those based in Singapore, were instrumental in the separation, as they were not comfortable with trading environment, which they thought could be stronger if the aforementioned challenges were addressed. After becoming a crown colony, Singapore's port started to thrive, and this attracted more people who came to live and work in the port city, and by 1870, the population had reached 100,000 people (Lambert, 2019). A map situating Singapore is represented in Figs. 1.1 and 1.2.

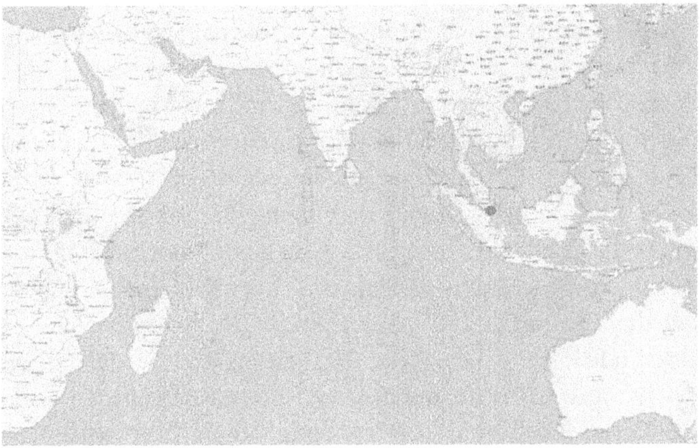

Source: © OpenStreetMap contributors.

Fig. 1.1: Location Map of Singapore.

Source: © OpenStreetMap contributors.

Fig. 1.2: Closeup Map of Singapore.

By 1942, Singapore was captured by the Japanese who ruled until 1945 when they were defeated by the British who again re-occupied Singapore. The following year (1946), the Straits Settlements that were part of crown colony were dissolved and Singapore was separated from Malaysia, and in 1954, a strong political movement, the *People's Action Party (PAP)*, was formed and through its agitation, Singapore's new constitution was inaugurated. Under this constitution, a new election was conducted in 1957 for members of the legislature and out of the available 51 seats being contested for, PAP won 43 seats. In 1959, LKY as the leader of the PAP led the country to self-governance and became the first prime minister. In 1963, to form a formidable regional block, Singapore joined the Federation of Malaysia, but the association was short-lived; thus, in 1965, Singapore pulled out to become an independent state (Lambert, 2019; Lim & Lee, 2016). In 1971, the last British military forces were withdrawn from Singapore (BBC News, 2018), and from then, with true independence and political and governance freedom, the country embarked on an expansive, robust and ambitious transformative journey.

FROM A THIRD WORLD TO FIRST WORLD COUNTRY

The true story of Singapore's success from third world to first world country is well captured in the growth of its gross domestic income, and also through the purchasing power parity per capita (OECD, 2011). In 1965, as explained by Menon (2015), Singapore's nominal GDP per capita stood at a low of US$516.5, a figure that was also experienced in countries like Mexico and South Africa (The World Bank, 2019a). By 1975, as showcased by Index Mundi (2019), following robust economic stimuli and attractive enabling environment, the country's GDP per capita had reached US$2,490. During this period, the country had maintained an average of over 12% economic growth even reaching a high of 38.9% growth rate in 1974. Another 10 years later to 1985, the GDP had risen by almost threefold to US$6,995, and by 1995, five years after LKY resigned as the prime minister, the GDP had risen to a high of US$24,936. As of 2016, the GDP per capita had reached over US$56,000 and from prediction, as noted by Menon (2015), it will reach a high of US$96,000 by 2040, a figure that will be higher than those of superpowers like the United States. Initially, these figures were dismal compared to those of its Asian counterparts like Japan (919.8) and Hong Kong, China (676.8) who were showing stronger economic growth than Singapore (The World Bank, 2019b). In 1975 as highlighted by Country Economy (2019b), the real GDP of Japan and that of Hong Kong were at US$521,542 million and US$10.048 million, respectively, while that of Singapore was only US$5,633 million. Today, the GDP per capita of Singapore stands at US$59,990 while that of Hong Kong is at US$48,517 and the Japan's is at US$39,228 (Country Economy, 2019a). The transformation of Singapore as depicted in these figures is so unique that no other country has managed to achieve a similar growth in a comparative timeframe.

TOWARDS ECONOMIC RESILIENCE

It also feels important to highlight the hardships experienced by Singapore before it experienced the above-mentioned triumphant changes. Before the British colony claimed the once silent, but active commercial hub that actively engaged with other ports in trade, there is little or no account of the island being poor, corrupted or polluted.

The only sound account is that of politics and power as different empires tried to control it to benefit from the commercial opportunities it exuded. But, when the British came in as explained by Buckley (1984) and started to openly display their thirst for power in the region, problems started to arise. First, this is attributed to the idea, which was supported by the local merchants of separating Singapore from the Straits Settlements that were under the British East India Company. The call was for Singapore to be recognised as an independent British crown colony, directly controlled by the British government. When this happened, the population of the port city expanded, leading to increased social problems; hence, people thought that falling out from the Strait would allow some of the problems to be addressed. But, unfortunately, the British were more concerned about their self-interests and those of the crown than those of locals (Bose, 2005).

Horton (2018) explains how the British left Singapore in an impoverished state characterised by poor housing, massive environmental pollution, poverty, low health standards and disjointed infrastructural development. This is partly due to the fact that Singapore played a significant role in the region in terms of trade and this attracted numerous confrontations and diplomatic incidents (Farrell, 2017). A notable confrontation involved the Japanese and the British Army in 1942, where the British were overpowered, but, in 1945, another war between these two groups broke and the Japanese were subdued. Such conflicts were precarious to the economy of the port city and had far-reaching consequences on locals who were also experiencing internal challenges like disagreements with the neighbouring Malaysia. The British who were fighting tooth and nail to keep control of the city are reported to have engaged in underhand practices like allowing widespread corruption to thrive and also omitting to address social challenges like housing, which Singaporeans were grappling with (Quah, 2018).

Austin (2009) explains that though Singapore had become a centre for international trade, the financial systems that were thriving in this city-state only served the interest of foreign exchange, with no trickling effect on the economy of the country. He accused the British of being self-centred in their failure to develop human capital sourced from the local economy or help implement special agricultural-oriented projects that would spur the well-being of the Singaporeans. Such trends were geared towards ensuring that Singaporeans would remain at the whims of their colonial masters. This was demonstrated by the poor liveability standards that are reflected in the poor housing schemes that were in place during the colonial era, and the wanton destruction of the environment, especially the Singapore River, which is showcased in Chapter 3.

The sorry state of the country is what prompted LKY to commit to uplift the liveability standards of his country by adopting and devising a series of best practices. Yew (2015b) shares that the status of the city-state when he took over leadership was dire such that all sectors of social, economic, political, environmental and technological spheres required an overhaul for them to support the state. To address them, LKY proposed the formulation of an Integrated Master Plan that captured both short-term and long-term development agenda. The purpose of the master plan led to promote innovation and flexibility in an integrated, balanced and dynamic environment. Areas targeted in the plan included providing decent and affordable housing for all Singaporeans, to ensure they live in an integrated community that capitalised on their cultural strength and advocate for moral and ethical practices. The plan also targeted the transportation sector where elaborated networks of roads, railways and airports were to be focused upon. LKY and his government also committed to promote practices like manufacturing and the creation of recreation centres that would improve the economy of the country (Han, 2017; Yew, 2015a).

The targets set in the master plans were not showcased for political mileage, though most of them have been achieved today and have exceeded expectations regarding their performance. For instance, under the auspice of the HDB, over 90% of locals own a house that optimises resource consumption. The houses are built in a compact neighbourhood and hence allow for the optimisation of scarce space that characterises Singapore. Another pointer to the attainment of the master plan is the ability of the country to provide safe and clean water to its citizens from within its borders. Initially, the country is said to have relied on water importation from Malaysia, but an innovative project undertaken by the Ministry of Water has allowed the country to clean used water under a high-tech initiative coined NEWater (Lee & Tan, 2016).

The country's education was also reformed to the standard where it is termed as amongst the best globally, and from it the country has managed to create for itself a highly talented and skilled workforce (OECD, 2011). Singapore now holds a great reputation in the field of science and manufacturing which are seen as the key drivers to the country's economic growth.

The health sector in Singapore was also developed to mirror those of the first world countries, and this has been credited for the rise in life expectancy (Akingba, Kaliappan, & Hamzah, 2018), where, as is depicted by the Department of Statistics Singapore (2019), the life expectancy is the second highest after Hong Kong. When all these social, political, economic and environmental advancement are summed together, they demonstrate the fulgurant rise of Singapore from a third world country to an attractive first world nation, and these happened in only one generation under the stewardship of LKY. The health sector is so advanced that the country is able to economically support an ageing population.

The sound leadership of LKY and his administration, accompanied by their vision of a better country and their dedication demonstrated, won the country the accolade of being the only country in the world to have transformed to first world country in one generation. As explained by Juma (2013), the country managed to achieve an all-rounded development. Singapore achieved success in social welfare and economic growth, while attaining admirable political maturity and promoting sound environmental practices, within the 50 years of rule of LKY. This is more pronounced noting that the country has no exploitable resource besides human capital (Chye, 2014). LKY's leadership style and skills, coupled with the firm belief in the possibility of transformation, led the country to graduate in both economic and liveability levels.

GEOGRAPHY AND POLITICAL CHALLENGES

The developmental potential of Singapore as a city benefited strongly from its geographical location. This is exemplified by the amount of interest that has been shown on its port ever since the era of ancient colonial empires like the Dutch Empire, British Empire and others like the Ottoman Empire and the Majapahit Empire. From the historical accounts narrated by Leyden and Raffles (1821) and Findlay and O'Rourke (2007), the protagonists of Southeast Asia were attracted to Singapore, and they were ready to fight for the control of goods in the region. This is better demonstrated by the amount of interest the British Empire had on the then port that they had to wage severe wars on their competing counterparts like the Dutch Empire and the Japanese, amongst others (Findlay & O'Rourke, 2007). But, the interest in having a strategically located port was supported by both a political and economic interest, often at the disregard of local populations. For Singapore, this is a different case – especially after the country gained self-rule, its leaders, led by LKY, demonstrated a strong will, vision and courage to unleash the best out of its geographical location. This began to take a more ubiquitous shape back in 1969 when Singapore became the first country in the entire Southeast Asia Strait to build a container terminal in Tanjong Pagar (Woo, 2017). At the time, the ultimate potential of this port may not have been clear, but turned out to be amongst the best, busiest and most connected port globally (Mody, 1997). C. Li (2018) explains that it directly connects more than 600 ports based in over 120 countries worldwide, and being closer to fast-growing economies like China, the traffic in this port city is enormous. She estimated that the port serves more than 130,000 ships every year, and as globalisation keeps gaining momentum, the traffic in this port keeps escalating. For instance, as showcased in the Singapore's statistics in 2018 (Department of Statistics Singapore, 2019), sea cargo increased from 593.3 to 627.7 million metric tonnes between 2016 and 2017, respectively. On the same container, throughput in the same period increased from 30.9 to 33.7 million twenty-foot Equivalent Units.

The geographical advantage of Singapore and the true investment in the maritime business has allowed the country to not only attract ship traffic, but an entire trading ecosystem range characterised by numerous financial institutions, shipping groups, ship broking firms, law firms inclined on maritime practices and many others (Maritime Port Authority Singapore, 2018). Through this maritime ecosystem, the country earns approximately 7% of its gross domestic product (GDP) with over 170,000 of its citizens and foreigners benefiting directly in terms of job opportunities. The strategic location and the potential of the port are also credited for attracting numerous foreign trade and investment, especially those banking on the openness and honesty of the government and its citizens at large. Currently the country's Foreign Direct Investment (FDI) standards stand at US$24.4 billion from US$19.1 billion recorded in 2018 and are expected to continue increasing (CEIC, 2019).

Besides the numerous geographical advantages that Singapore prides, it has experienced thorny challenges which are imprinted in its history. The main challenge that Singapore still continues to encounter is the supply of sufficient, clean energy for both domestic, agricultural and industrial use (Ministry of Trade and Industry Singapore, 2018). For this reason, besides the challenge of insufficiency as showcased by Temasek (2018), transitioning to cleaner energy production is a main hurdle that the country seeks to address. The commitment by Singapore's government is to supplement its reliance on fossil fuel by adopting green energy technologies and to eventually reduce emission intensity by 36% by 2030. By doing this, noting the potential of alternative energy especially solar and wind power, the country can manage to lower the cost of electricity, as well as save on other urban fabrics that are directly or indirectly impacted by the use of energy from non-renewable sources.

Another aspect of importance is the provision of clean water to power both domestic, commercial and industrial activities, highlighting that most of the critical sectors of the country heavily rely on water. For instance, Singapore's National Water Agency explains that the water demand in Singapore lies beyond 430 million gallons each day, and as growth in different sectors continue, the water demand will continue to increase (PUB, 2018). Therefore, despite investing heavily in water supply diversification, like the reclamation of used water through its NEWater project, desalination and spirited efforts in using local catchments (Lee & Tan, 2016), the country still relies on Malaysia for most of its water needs. Long (2001) highlights how this reliance has been rocky, as Malaysia has in some instances capitalised on this weakness of Singapore to use water as a political tradeoff between the two economies.

Besides the two challenges, Singapore was also struggling with its small size, which was seen as a disadvantage especially due to the increasing population. This increased the demand for services and resources, which are, as mentioned previously, scarce in Singapore (Ludher, 2016; Tortajada & Hongzhou, 2016). The available space was not enough for collectively providing housing, infrastructural development and for extensive agricultural practices that would ensure food security. For this reason, Singapore had, at all cost, to rely on its neighbour for food importation. The costs for such importation were increasing consistently, since, as En (2017) explains, they were influenced by global and regional economic situations. For this reason, the country was struggling in this front, and this through the leadership of LKY prompted robust initiatives such as the partnering with neighbouring countries like China for food production. A case in point is the aforementioned SSJFZ that is based in China (AFJ Editor, 2010). In addition, the Singaporean leadership opted to extend its importation base, which, to date, relies on over 170 countries across the globe. The idea of reclaiming part of the waterfront to increase the habitable land was also borne out of such challenges, and as Rut and Davies (2018) highlight, the government succeeded in availing underutilised spaces for gardening to locals.

INFRASTRUCTURE BUILDING

When Singapore gained independence in 1965, the country's infrastructural developments were disjointed, insufficient and in dire need for substantial renovation. There was not much to build from and the government of LKY did everything almost from scratch. To put this in perspective, by the time the British allowed Singapore to self-rule, the country was facing a severe housing crisis, topped by massive pollution and limited resources. According to Yuen (2007), in 1947, a report by the British Housing Committee cited Singapore to have the world's worst slums that were overly overpopulated. Jha (2018) reports that in 1959 only 9% of Singaporeans owned a house. This number has exponentially increased to over 90%, where today Singaporeans comfortably live in their own houses built by the government-owned HDB. He reports that before independence, 72% Singaporeans were scattered within the 80 square kilometres that was considered as the main central business district (CBD), but in the modern days, most of the population is distributed in over 23 self-contained housing development built in the country's new towns that extend beyond the city's waterfront. The advantage of such housing planning is that a compact neighbourhood allows for smart use of urban resources like energy and water (Gaigné, Riou, & Thisse, 2012), which, as noted above, are scarce resources (Shatkin, 2014). An example of the aforementioned housing projects is shown in Fig. 1.3.

Mauro (2018) expresses that the housing program would not have been such a success without the notable investment in transportation infrastructure interlinking the new satellite towns to the CBD. Singapore heeded this call by ensuring first-class expressways and rail system, almost simultaneously with the housing program. Besides internal mobility which was essential, Singapore is painted as one with first-class

Source: Image by Steven HWG.

Fig. 1.3: Housing Development Projects by the Singaporean Government.

infrastructure connecting the city-state with the rest of the world (Government of Singapore, 2018). In particular, the port, which has remained one of Singapore's critical resources, has been under massive development, and as noted above, it has the capacity to handle large volumes of cargo into and from over 600 ports worldwide. The port (Fig. 1.4) is linked to a complex, high class web of road and railway network that allows quick, safe and free movement of goods to and from the manufacturing zones, markets and port.

Source: Image by Chuttersnap.

Fig. 1.4: Singapore Port.

According to the Ministry of Transport (2019), the road network in Singapore occupies approximately 12% of the total habitable land surface. This translates into approximately 86.28 km^2 of its 719 km^2 of surface area, where most of the country's area is attributed to road transportation. While noting that Singapore's transportation is fluid and efficient, it is also important to highlight the considerable amount of space occupied by public transportation. This has a lot to do with the increasing population and lack of available exploitable space for development. This compactness can be seen in Fig. 1.5, showing how the urban areas are made to accommodate for vehicular flows.

Lahiri (2017) explains that by 2030, the country's target is to have a car growth rate of 0% to retain the expansion of the road network cover – currently standing at 12% – to a minimum. This figure can be projected to reach 19% by including both land mobility and the airport expansion projects to cater for an increasing projected population of 7 million by 2030. This limited expansion in road network,

Source: Image by Fahrul Azmi.

Fig. 1.5: Road Infrastructure at Lau Pa Sat in Singapore.

even though a projected increase in population is noted, supports the usage and investment in public transportation, rendering the country even more accessible, but at the same time less dependent on private vehicular transportation. The road network also connects to the three main airports, playing a critical role in ensuring the movement of people and goods to and from all over the world. To cap the serious investment in the air sector, Changi airport, the largest airport in Singapore, was ranked by Skytrax as the best airport in the world, a position it has held for 7 consecutive years (Gilchrist, 2019). The airport recently opened a new terminal (Figs. 1.6 and 1.7) which has gained international fame due to its innovation in design.

Other notable infrastructure projects that have received serious attention since the independence of Singapore include the energy production, the Singapore River and some substantial cleaning programs (Choong, 2018). As explained above, energy production has been key to the country as it serves a pivotal role in the manufacturing sector, a sector on which LKY's government had pegged a lot of hope in

Source: Image by Andy Yeo.

Fig. 1.6: Terminal 3 at the Changi International Airport.

Source: Image by Andy Yeo.

Fig. 1.7: Terminal 3 at the Changi International Airport.

achieving the economic growth of the country. The river-cleaning programs were very successful, especially in the Singapore River and Kallang Basin. The latter was transformed from a filthy, badly polluted river into a clean, attractive and serene environment characterised of trees and attractive vegetation (McNeur, 2017; Ministry of the Environment and Water Resources, 2018; Newman, 2010) promoting human and environmental dimensions (Tortajada & Joshi, 2014). Ooi (2005) shares how LKY perceived the cleaning exercise as part of the economic growth and development strategy, and as it stands today, his vision is still valid as the water bodies give Singapore a liveability edge. Y. K. Joshi, C. Tortajada, and A. K. Biswas (2012) explain how the leadership style of LKY ensured the cleaning exercise was complete within a record period of only 10 years (1977–1986). The cleaning exercise was carried out in tandem with other infrastructural development programs, such that by 1990, when LKY resigned from office, Singapore had achieved an accolade in the development arena as one that moved the country from a third world to a first economy in only one generation.

The massive infrastructural development projects in Singapore came at a certain economic cost, especially for a young, poor and resource constrained country. Even though the situation, under LKY, Singapore exploited a strategy that is deemed as one of the most potent and successful. This is captured in a highlight by Alam (2015) who highlights how the then prime minister emphasised the need for foreign investments in Singapore. According to him, LKY understood that creating an enabling environment that was free from corruption that offered a mix of tax incentives and allowed free trade would attract rich, reputable and experienced corporations, financial institutions and private investors into the country. And true to this, by ensuring that sound governance and ethical practices anchored in government policies were in place, Singapore attracted numerous FDIs. These, besides creating massive job opportunities for the increasing population, also allowed a substantial flow of financial resources in the country that aided in the undertaking of the aforementioned projects.

With considerable amount of finances and correct skill set from foreign experts that had migrated to Singapore to take advantage of the transforming economy, most of the projects as showcased above were delivered in a record time. Rana and

Lee (2015) posit that, to date, the innovative strategy of relying on the export-oriented development model adopted by LKY is seen as one of the most potent that developing economies should endeavour to mimic. Nevertheless, LKY also favoured the knowledge-based development model where industries skilled in such areas were given attractive incentives into Singapore. When such engagement led to investments in specific and specialised sectors, Singapore besides benefiting from job opportunities and financial flows also enjoyed knowledge transfer and capacity building, leading to the setting up of smaller local companies that had the potential to compete on the same platform.

CONCLUSION

The chapter thus introduces Singapore as a city-state and surveys its historical background around how the British colonisation led to further accentuate its geopolitical and economic significance through a heavy investment in port activities. Singapore remained relatively poor during the colonial era, where most of its riches were exported to the British empire, and it was only after independence, through the reign of Lee Kuan Yew, that Singapore experienced an economic success that was delivered at such a rapid pace, that its model is now being studied in many parts of the world and being replicated in the hopes that this will bring prosperity in varying contexts. The leadership adopted during the post-independence era was also key in unlocking its developmental infrastructures that the country enjoys today.

2

SINGAPORE'S GOVERNANCE STYLE AND URBAN PLANNING

INTRODUCTION

There are varying stories describing the amazing journey of Singapore and how it transformed from a meagre economy to the league of world economic giants in a record time of only one generation. But such stories are seen to always revolve around Lee Kuan Yew (LKY). From a myriad of literature and discussions in different forums, LKY has been referred to as the founding father of Singapore (Ramdas, 2015), despite Singapore's history dating many years back. Throughout this prior history, it was noted that those who sat at the leadership seat of the then Island city were only driven by egocentric and selfish interests (Borschberg, 2010). People were fascinated by the potential of Singapore as a strategic trading point and were thus willing to even engage in war to defend or capture it as part of their own trading purposes. Such included the Dutch and British Empires and the Japanese, amongst others (Leyden & Raffles, 1821). Of these, Sir Thomas Raffles, who set foot in Singapore in 1819, is the only one known to set commitments towards development to give Singapore a glimpse of hope of becoming a significant trading hub in the Southeast Asia and with the rest of the world (Ting, 2019). But, most of the things he and the British government he worked for was oriented towards the port, and this was not innocent as it was aimed for the welfare of their government and for the betterment of companies in colonial home. Despite the fact that the island was prospering on many grounds, locals were living in slums with poor sanitation, little or no clean water, massive unemployment, low-quality health services, poor education systems and massive pollution amongst other problems (Hean, 2017; Turnbull, 2009).

When LKY took over the reins of government, he was crowded with a myriad of challenges to prove that Singapore could manage to self-rule despite lack of natural resources and their wanting status. He also wanted to prove that the country did not have to be merged with its neighbours like it was the case with Melaka and Malaysia, under a common name of 'Federation of Malaysia', which did not last long after the relationship between these countries started to waver (Haackle, 2005). Amongst those issues that resulted into the collapse of the union include corruption, unfair taxation and underhand trading practices, amongst others, that LKY did not want to be repeated in the independent Singapore. For this reason, his leadership and governance style was characterised of openness, honest, commitment, adherence to the rule of law, zero tolerance to corruption, society oriented and altogether visionary, as depicted in numerous literature (Choi, 2018; Civil Service College & Centre for Liveable Cities, 2014; Juma, 2013; Rana & Lee, 2015). But, despite the numerous and tangible developmental advancements and benefits that Singapore leaped from the leadership of LKY, some critics argue that his leadership style was autocratic and very far from being democratic (Meadows, 1988).

Choi (2018) supports that mostly, such critics use the same arguments as one used in western countries, but, despite their view, highlighted issues like lack of public consultation – that they use to dismiss LKY's leadership – may be far-fetched. This is particularly true noting that LKY was noted as merely driving a master plan that was borne out of wide scope participation from the government official appointed in different capacities (Global is Asian, 2017). The style of governance does, however, support differences from western definitions of democracy, as this chapter introduces.

Interestingly, from inside Singapore, this autocratic style is not often viewed as such, but is agreeably perceived as a 'strong leadership' style, and mostly commended. This is so because it is understood that it is this strong leadership style that is credited to the success of Singapore and to its infrastructural development that have earned it the privilege of being cited as the case study to many cities and countries. A pointer to this is the establishment of the Housing Development Board and Economic Development Board that were established by LKY to spearhead housing projects, which, as seen in Fig. 2.1, are now testament to the success those brought about. Additionally, those allowed the expanding of economic bases through industrialisation and manufacturing to increase employment opportunities (Chye, 2014). These two boards that were created in 1960 and 1961, respectively, and have steered the country to immense success in their respective fields have been headed by different committees since then, hence, supporting that LKY was just a motivating figure, who through his charisma and passion motivated others to surmount challenges that may have deemed daunting to accomplish (Chen, 2016; Sigdyal, 2018).

The role of governance and urban development is strongly linked (Allam & Newman, 2018b), and this has often been demonstrated in various contexts. This is particularly

Source: Image by John T.

Fig. 2.1: Housing Complex in Singapore.

influenced by the type of leader and leadership style adopted. However, literature indicates that most people, especially those drawn from the western countries and other parts of the world, seem to favour democratic and participatory planning.

Nevertheless, Singapore stands as an interesting case as it has shown otherwise and has successfully emerged economically stronger and more resilient (Choi, 2018; Civil Service College & Centre for Liveable Cities, 2014; Crinson, 2017; Lim & Lee, 2016; Mauro, 2018; Tan, Chuah, & Luu, 2018; UN Environment, 2016). This argument, though, does not dismiss the need for participatory planning and an all-inclusive governance structures, but one need to understand the underlying issues that led to LKY adopting this unique leadership style. In the foremost, the unique case of Singapore is tied to its history, demonstrating a country that underwent numerous and tantalising promises from different leaders who ultimately turned out to have been selfishly driven by greed, and not for the welfare of the population at large. It was therefore in a downgraded state and needed speedy transformation to at least support a humane environment for people to live and thrive.

The demographic composition of the country was also an important factor (Stuart, 2015). Unlike most countries, Buckley (1984) highlights that the multiplicity of cultures in Singapore – where each group had their own view of things and varied ambitions – could have led to various levels of resistance to change and derail the vision and the speed at which LKY wanted his country to transform. However, LKY's strategy was to ensure that a multi-cultural society prevails and above many other reasons that can be cited in the case of Singapore and why LKY's leadership and governance style worked has mainly to do with the trust and belief that the people of Singapore placed in him as a person and as a leader (Austin, 2009; Ministry of Trade and Industry Singapore, 2018; Yew, 2015b). He, unlike his successors, had held different leadership portfolios, from prime minister, minister and government advisor for 56 consecutive years unopposed, and throughout that period, the country had numerous, tangible successes to show from his efforts (Burton, Montagnon, Brown, & Grant, 2015; GovTech Singapore, 2018; Ong, 2017).

DWELLING INTO SINGAPORE'S AUTOCRATIC GOVERNANCE STYLE

There is a wide difference between autocracy and democracy, and this can better be demonstrated in the impact such governance styles have had in different economies. In many western countries, and the United States, the latter type of leadership is preferred and emphasised. But, in Singapore and some others, the former is adhered to, and has worked quite well as demonstrated in Chapter 1. In plain language, democracy means that the power lies in the hands of the people, while in autocracy power is deemed to be centred on an individual. In a democracy, as Demir (2017) opines, there are a number of issues pertaining to independence that must stand. These include freedom of thought and opinion, freedom of expression and impartiality and finally, free, fair and independent justice system. Fishkin (2018) argues that deviation from any of these perceived freedoms is seen as a threat to democracy, and such threats are said to be protested, at least in most countries, through set mechanisms that are anchored in the constitutions of such countries. And, to be fair, this kind of governance has been instrumental in advancing the right of people in different countries, and has held governments in checks. But, from observation as noted by Andelman (2018) on different economies in the modern days, governments such in United States under President Trump are seen as though they are slowly shifting to autocracy, and such trends need to be evaluated. On the other hand, in an autocratic governance style, most of the rights enjoyed in democracies are limited, and in most cases those in power have the power to instigate their actions as being the sole right way of operating and those with dissenting opinions are not tolerated (Tullock, 1987). Such rights like the freedom of media are greatly curtailed, and the reporting organs are deemed to be controlled by the state or those in power.

In Singapore, for varied justifiable reasons, the autocratic leadership style seems to be the most preferred, and the most suited for the country, though there has been some calls, especially from the west and global north for the country to embrace a more democratically oriented leadership style.

Some quarters even equate autocracy and the leadership style of LKY to dictatorship, but, with an equal measure, such perceptions are largely disregarded as dictatorship is usually equated to negative economic repercussions as opposed as to that seen in Singapore. Indeed, from the literature above and based on the general observations of the country and the progress that has been achieved over the 56 years that LKY was in power, notable successes were achieved. Kwang et al. (2011) describe LKY as a charismatic leader who was flexible and adaptable to the changing dynamic, especially in regard to economic development. He has also been described as a coach and a motivator who cherished teamwork and was also a convincing negotiator, strategic planner and executor who was ready to put the society ahead of his personal interest. For instance, it was said that through his rein, he prevented the internationalisation of the Singapore dollar by limiting the operations of foreign banks, which would have disrupted the domestic financial market (Srinivasan & Negandhi, 1985). Through his mentorship, his cabinet and those in different ranks of government were able to convince numerous companies and corporations to invest in Singapore, and as of 2018, according to Singapore Business (2018), there are over 46% multinational corporations with offices or their headquarters in the country, ahead of Hong Kong which hosts 37% of all multinationals in Asia. Such reputable investments could not have been achieved if the country was under a dictatorial leader as we know them. Another factor brought forth by literature showcasing that LKY was not a dictator nor the type of autocratic leader he was accused of is his decision to gracefully resign from the prime minister's position for others to continue serving the country (Han, Fernandez, & Tan, 1998; Quah, 2018).

In his autobiography (Yew, 2000), LKY candidly stated that he would not like his elder son, Lee Hsien Loong, to succeed him after he left the prime minister's office since it would be perceived as nepotism, even if his son deserved the post. According to him, the position of the office should be earned through popular vote and not having someone hand-picked without consideration or merit. Despite these selfless words for the love of country, Loong went to become the prime minister, but 14 years after his father had resigned. Between that period, the position was held by Prime Minister Goh Chok Tong who, like LKY, led the country through successful periods. At no time did the country's economy waver due to change of guard in the position of the prime minister. Mr Loong, LKY's son, as reported by Berge (2004), had every right and qualification to take up the position of prime minister as he had the right academic qualifications and also had gained leadership experience, especially during his years in the Singapore armed forces where he held different positions, with the highest being a brigadier general. He had also held the position of deputy prime minister under Prime Minister Goh. Similarly, besides his qualification, the people of Singapore had confidence in him and desired that he continued his father's impeccable legacy.

Despite its applicability in Singapore, the autocratic type of leadership as described above is not always successful. Its characteristic of having an individual yielding massive power, either expressed from the constitution or from adopting such style of leadership, has been abused in numerous regions, especially in developing countries just as well as in middle-income countries (Lee, 2018). Countries such as Zimbabwe, Uganda, Vietnam, Ethiopia, Iran and Sudan amongst many others have been under autocratic leaders, who, unlike LKY, have been on a self-serving mission while a majority of their citizens wallow in poverty, hardships, poor education systems, poor health services and brutality from security forces amongst many other issues (Trindade & Costa, 2011). In Zimbabwe, for instance, the former head of state – President Robert Mugabe – remained in the topmost political position for more than 30 years and left the country with dire economic and social problems (Chappell, 2017). He is accused of utilising every available opportunity to intimidate the opposition, foreigners and the media and was forced out of office by the military after increasing popular pressure fuelled by a general feeling of required political change (Winter, 2017). Similar predicaments were shared in Sudan, where the authoritarian leadership of Al Bashir was cut short through peaceful demonstrations by the citizens who had so many issues to accuse him of (BBC News, 2019).

LKY's governance styles were not without some pitfalls. Some notable ones include his hard stance on the media and his uncompromising reaction to those who opposed him, whom, as explained by C. Tan (2015), were mostly jailed without trial. To him, these were distractors threatening to his ambition and vision for a better Singapore. Nevertheless, Singaporeans were noted to be continually satisfied with what he achieved for the country, especially through the direct impact the rapid transformation brought into their lives. Specifically, they were grateful for the employment opportunities, the competitive wages they received – courtesy of government intervention – and the ever-increasing GDP per capita, amongst many other benefits.

LEADERSHIP AND URBAN DEVELOPMENT

Unlike democratic leadership style that is preferred in many countries and is promoted for its emphasis on its ability to allow both top-down and bottom-up consultation, strong autocratic leadership has the potential to yield resounding results. As has been demonstrated by LKY, who is said to have practiced a relatively 'soft' authoritarian style, this type of leadership provides a variety of freedom enjoyed by the leader to pursue specific ideas that can yield tangle results in a record time. In Singapore, where numerous examples can be cited, every single project that LKY set his mind and effort was ultimately successful despite its magnitude and scale. For instance, the housing project that was initiated after he created the Housing and Development Board (HDB) was so successful that other countries are now trying to mimic it (Jha, 2018; Yuen, 2007). In reality, if such a project was to be undertaken in a democratically led economy, it could have derailed and would not achieve the same outcomes. To put this in perspective, Singapore is amongst the leading countries in terms of home ownership, ahead of large economies with massive resources and many years of independence at their disposal. According to a report by Trading Economics (2019), Singapore is the third globally in terms of home ownership, behind Mauritius and Romania that have 99.1% and 96.8% ownership, respectively, while Singapore is at 91% and it achieved this in about 50 years of independence. The United Kingdom, who colonised Singapore in the past, is at 65% (Trading Economics, 2019). This is a very ironic situation noting that the United Kingdom was initially richer in resources which were sourced both locally and from its colonies.

The achievement of diverse projects as seen in Singapore is an outright vindication of its autocratic leadership style, but this can only be sensibly argued to be possible when this is applied by the right person, and not self-serving authoritarian leaders. This governance style can be paramount in saving time while allowing for optimal uses of resources, both in human capital, financials and raw materials. But at the same time, the beneficiaries of projects need to be informed of the decisions and the way forward; hence, they are better aware of their responsibilities and that of the government. This was demonstrated by LKY as explained by Hyman (2015) who noted that LKY would regularly, and in a layman language, explain the government's future plans and what it entailed to his Singaporean counterparts via the radio, and this way kept them on tab on what the government intended to achieve. One characteristic that enabled him to convince his constituents about the government agendas of which they were seldom part of was his approach of honesty and the foundation of corruption-free environment that he had established (Jones, 2016; Quah, 2007). Participatory planning, which is often seen at the complete end of the spectrum to the dimensions supported by autocratic leadership, can be time-consuming and often contested as it takes a lot of time, often more time than political mandates. Thus, urban development fails to be achieved at desired speeds, or even at the speed at which infrastructures are needed.

This is true in many countries, even in those that are said to experience real democracy, as leaders are often seen to capitalise on projects running beyond their terms in office to seek fresh mandates in pretence that they would like to complete the project they initiated. These assertions are supported by

Hibbing and Theiss-Morse (2002) that are of the opinion that most citizens in any republic are not well informed about public policies; hence, it would take time to explain to them and gain their approval; hence this is a time-consuming exercise. To avoid the trap of politicians taking advantage of such ignorance to advance their interests and earn new political mandates, the authors propose an authoritarianism style of leadership where technocrats and experts are appointed to run government projects, thus avoiding unending debates and unnecessary compromises while having the country run by informed and learned leaders. Irvin and Stansbury (2004) explain that a participatory approach cannot be made for every project or process and is better implemented when the citizenry in question is not passive. They support the notion that public participation, in most cases, leads to extra social and economic costs that make projects undertaken under the regime expensive, time-consuming and not always the best suited for the issue at hand (Leong, 2000).

The dire state at which Singapore was at the time of independence could not afford it the luxury of engaging in endless debates on what projects were to be implemented and which ones were to wait. LKY, through a book authored by Kwang et al. (2011), explained that the country was still vulnerable from issues like external enemies, including its neighbours who would have wished it to fail. Also, due to limited resources and poorly managed public sector, there needed stern decision to be taken (Doner, Ritchie, & Slater, 2005). For this reason, the guided authoritarian style that LKY adopted was maybe the best for Singapore, and the achievements were evident. Even critics like Morgenbesser (2016) who does not approve of the Singaporean autocratic approach to leadership agree that, at least, through LKY's style, Singapore managed to make progress, where many and even most democracies would have failed terribly. Others like Guriev and Treisman (2015) who view LKY as a benevolent dictator also marvel as to how Singapore had managed to achieve so much under the autocratic leadership, and argue that at least Singapore and other countries like Russia – that are seen to be under autocratic leadership – should be democracies. But, as is showcased in the next section, Singapore though under an autocratic leader did not experience a total shut down in terms of public participation. At least, through the parliament, and the different government agencies, sound and concrete decisions that were not as much influenced or under the patronage of LKY were made. Those decisions contributed greatly to the liveability standards of the Singaporeans. Such include the housing project, which, though directed by LKY, was implemented by HDB and the same was also responsible for the distribution of the houses to the locals (Chia, Li, & Yang, 2017; Rocha, 2011).

PUBLIC PARTICIPATORY PLANNING IN SINGAPORE

Even though Singapore was led by an autocratic regime, there have been notable emphases on participatory planning, where the views of citizens and residents have in most cases been sought and given due attention, especially in planning projects. A case in point is the framing of the country's master plan that was first initiated in 1952 and which has been reviewed to accommodate for arising public concerns. It was noted that firstly, the review was done every two years, but after 2008 this was performed every subsequent five years. Throughout this process, the views of the citizens were ceremoniously considered. Liu (2016) acknowledges that the master plan has remained relevant and effective due to a close collaboration between politicians, who represent the general society and professionals from different backgrounds. Another project that demonstrates a close collaboration between government and citizens is the formulation of the *Centre for Liveable Cities (CLC)*, which in turn helped in the formulation of the *Liveability Framework* that was to guide the country in achieving better liveability standards for the citizens. Chye (2014) explains that it was achieved due to quality inputs from professionals, public organisations and civil groups based in Singapore, and also, it is said that the CLC borrowed some insights from best practices of other cities across the globe. Again, the fourth principle in the Singapore liveability framework is captured in detail and explains the benefits of involving the community as stakeholders in various stages of the planning process. Wong and Fook (2016) further support that the public has been engaged in land-use planning, and for this reason the country has managed to optimise its limited space, such that it has been able to accommodate residential neighbourhoods, transportation networks, green spaces and many other infrastructures that have made Singapore the city-state it is today (Newman, 2010; OECD, 2011; Tan et al., 2018). Heng (2016) explains that, to some outsiders, the nature of public participation in Singapore may be confusing since the country organises a unique brand of public engagement which is somehow controlled by the organising agency but works well for the country.

After successfully coming up with the said documents (Master Plan and the Liveability Framework) focused on guiding Singapore into higher heights of success in various spheres, the country no longer stands as coy island state still facing the harsh consequences of manipulation by the British and by its neighbours. It now dines on the same table with the world's most advanced economies (Chan, 2018; IMD, 2019; Yong, 2019), and this is attested by Singaporeans, who for over five decades, as described by Guo (2018), have legitimised the leadership and development approach of the People's Action Party (PAP). Since independence, they have voted in the PAP government, and in return, the ruling regime has transformed key spheres of the country. Guo (2018) highlights the satisfaction that most Singaporeans have expressed following numerous projects like the housing program that have allowed over 91% of the citizens to own low-cost, but quality houses. He showcased how such projects have even become more successful through the participation of the residence where they proposed the local issues that they would love the government, through the HDB to implement to facilitate more liveability standards. Through such engagement, facilitated by the National University of Singapore (NUS) in 2012, the locals were able to propose neighbourhood designs that would enhance their bonding and social cohesion, and through this, Singapore now has numerous green spaces, ubiquitous in every city, to supplement the liveability standards enjoyed by the average Singaporean.

Gleason (2018) dwells on the progressive and robust education system that Singaporeans undergo, thus, equipping them to take up responsibilities commensurate to the standard of the challenges facing their vast growing economy. According to Yeo (2016), the changes in the education systems are borne from public input where these are implemented in the form of education policy through the Ministry of Education (MOE) of Singapore. For instance, in the 1990s, the public agitated for change in the education system to include innovation and technology in school curriculums, and in 1997, a master plan for IT in education was launched. In the water sector, Cecilia Tortajada and Yugal K. Joshi (2013a) explain how the citizens have been vocal in shaping the water policies. This eventually led to water policies that Singapore ranked amongst water resilient countries, despite its initial problems that prompted it to sign a water import agreement with Malaysia. Through the participation of different stakeholders who proposed a myriad of initiatives and campaigns for conservation of water, Y. S. Tan, Lee, and Tan (2009) and C. Tortajada and Y. K. Joshi (2013b) showcase that Singapore has achieved water resiliency. For this reason, Ching (2015) explains how after the reforms Singaporeans feel gratified by the availability of clean drinking water, which was not the case 50 years ago (Lee & Tan, 2016; Long, 2001; Luan, 2010). The clean water bodies and the pollution-free environment are seen as a blessing to the Singaporeans, and Khew (2015) explains the commitment of the citizens to maintain these standards – initiated by their government.

In today's modern days, though it is evident that the freedom of the press is not well observed in Singapore, participatory processes like have been previously showcased had been observed. Specifically, with the advent of technology, the citizens have been vocal and active in participating, especially via online platforms (Ho, 2013). Baum, Yigitcanlar, Mahizhnan, and Andiappan (2017) explain that the government and its agencies regularly invite the public to participate, by commenting on development projects, and their views are evaluated and those relevant are considered when project applications are being processed. Feraco and Med and those r2014) explain that the utilisation of ICT, especially in communication, has enhanced government interaction with local populations, and through this the government is able to capture public data which help in optimising the planning process, with the ultimate aim to provide a better and mode integrated project. In times where has increasing importance in transforming cities (Allam, 2019, 2020a, 2020b, 2020c, 2020d, 2020e, 2020f), this is treated and guarded jealously especially through the various cyber security policies, and through it, Singapore is almost at the apex of becoming amongst the best smart and data-driven cities.

CONCLUSION

This chapter highlights that the success of Singapore in overcoming post-independence challenges and in providing for critical infrastructures was only due to the governance style chosen by Lee Kuan Yew and by its acceptance of the population. It is noted that no other country has managed to achieve such success in such a short time span and those projects would not have been effectively delivered if the then prime minister did not have the flexibility in driving those projects the way he did. However, it was seen that even though a strong governance style was adopted, the use of participatory planning has always been a key dimension of project implementation, and was used as a tool to devise better integrated projects to ensure the fair ownership of the people.

3

SEEKING LIVEABILITY THROUGH THE SINGAPORE MODEL

INTRODUCTION

The concept of liveability encompasses a number of indispensable and fundamental aspects, including economy, environment, governance, living and mobility (Allam, 2012; Southworth, 2016). On the same, Herrman and Lewis (2017) added to these by including the dimensions of economic competitiveness and social welfare. The availability of building spaces, transportation infrastructure, social inclusion and resources such as water, food, energy and raw materials are also seen as key factors that make a city liveable (Allam, 2018; Hunt, Makana, Jefferson, & Rogers, 2016). While the world is facing the impacts of climate change, researchers add that in the future, liveability of a city will be determined by other human needs like resource security, well-being and sustainability agendas. While there are numerous key dimensions that can improve quality of life, the increasing migration to cities adds to density, which causes a range of issues like environmental degradation, traffic congestion, lack of clean water, inefficient and delayed service delivery, and noise and waste management hiccoughs, amongst many others (Howley, Scott, & Redmond, 2009), hence impacting on the physical as well as the psychological well-being of inhabitants. While there are a range of technological solutions, the challenge remains as to how to tackle those issues in innovative ways so as to encourage a truly comfortable urban fabric (Allam & Dhunny, 2019; Clements-Croome, Marson, Yang, & Airaksinen, 2017).

Norouzian-Maleki, Bell, Hosseini, Faizi, and Bahram (2018) and Allam, Tegally, and Thondoo (2019) support that ultimately liveability is the extent to which different urban aspects work together to increase quality of life, which is measurable. The Economist Intelligence Unit (2018), dedicated in surveying the global liveability index, notes numerous factors with an emphasis on safety and stability. In an era of common resource flows coupled with urbanisation, these rankings can play a significant role in the decision for business and human migration.

In different urban spheres like planning, sustainability and resilience, amongst others, it seen that liveability concepts can allow cities to attract capital, innovators, wealthy individuals and talents, amongst others (Allam, 2018; Clements-Croome et al., 2017; Giap, Thye, & Aw, 2014; Norouzian-Maleki et al., 2018). These resources in turn attract further developments and creativity, which impact on economic competitiveness, political influence both locally and globally, the ability to attract socio-cultural innovativeness and the ability to attract visitors. The varying mix of dimensions can thus act as a catalyst to influence and guide public policies and national lifestyles (S. Teo, 2014). In particular, the increased use of technology in city governance and management are in the forefront in promoting this (Allam, Dhunny, Siew, &

Jones, 2018; Barns, Cosgrave, Acuto, & Mcneill, 2016). An emerging concept is that of the smart city which has been adopted worldwide (Silva, Khan, & Han, 2018; Ubaldi, 2013), and can be seen as a driver for economic opportunities (Allam et al., 2018; Allam & Newman, 2018a). However, an interesting smart city concept is that of the Mauritius model (Allam & Newman, 2018b; BOI, 2015), which focuses on the three dimensions of work, live and play, which ensures an inclusive planning concept and which was incidentally from Singapore model (URA, 2006).

The city of Singapore has had a good standing in global liveability index, particularly compared to its peers in Asia-Pacific. In 2018, it was ranked 11th in the region in terms of liveability, and scored the highest globally in terms of improvement in public healthcare category (The Economist Intelligence Unit, 2018). This ranking was achieved even though it is the third most densely populated city in the world. Its transformation agenda is strictly tied to the city governance and the underlying political environment. One of the most outstanding characters of Singapore is the intolerance to corruption, which is a major thorn to any development agenda. A recent report by Heritage (Heritage, 2018) credits Singapore's success to the improved government integrity, labour freedom and well-structured property rights. Even though the model of governance is disputed by many (Chong & Chan, 2017; Verweij & Pelizzo, 2009), unemployment in 2018 is only at 1.8%, and is an improvement from 2.2% as reported in 2017 (CIA, 2018). The consumer purchasing index (CPI), which is captured by the inflation rate, stood at −0.5%, and the per capita income as captured by Heritage is at US$87,855.

The attractiveness of Singapore amongst the most liveable cities globally is not incidental but is a result of a well-thought strategy. The country prides itself as one of the highly developed free-market economies in its region and globally. According to Heritage (2018), it scored 88.8% – second after Hong Kong. The ranking is partly influenced by the political stability that the country enjoys and the trade path of liberalisation and globalisation. A combination of these environments has allowed it to attract substantial Foreign Direct Investment (FDI) totalling to over US$61.6 billion (UNCTAD, 2018). In 2010, the literacy level in the country was over 98% and the education system is seen as one of the most competitive in the world (OECD, 2011). The labour force thereof contributes in making the country one of the major manufacturers of goods and services that are in demand worldwide. The country facilitates these exportation through its port, which is touted as one of the world's largest (CIA, 2018). Besides the port activities, Singapore has one of the most modern, progressive and sustainable transportation infrastructures in the world rendering an intricate connected network (Government of Singapore, 2018).

Those numerous achievements contributed to building a sustainable environment. However, while those may seem disparate achievements, those emerged from a liveability framework adopted early on during the emergence of its political independence by the Government of Singapore. This chapter reviews the Singapore's Liveability Framework and dwells into key elements of the urban environment contributing to its success.

THE SINGAPORE'S LIVEABILITY FRAMEWORK

One of the commendable features of Singapore is that development agendas are steered by the government from their conception stage till completion. This was ensured by Lee Kuan Yew during his reign as Prime Minister, since independence from being a British colony, and it has been consistent since then (Tan, Chuah, & Luu, 2018). Since 2008, development, research and policy agendas were compounded under one body named Centre for Liveable Cities (CLC), under the aegis of the Ministry of National Development that encapsulates the input of professionals, public organisations, civic grounds and inhabitants. This body refined the existing frameworks in Singapore and devised a new model that captures the concept of liveability and sustainability, which have become the corner stone of Singapore's success: the Singapore's Liveability Framework. This reflected on the cultural diversity presented in different parts of the country in a context with geographical limitation (Civil Service College & Centre for Liveable Cities, 2014), where only an inhabitable space of approximately 710 km^2 is present, which was increased through sea reclamation from the original size of 581 km^2. The habitable space is currently occupied by over 5.7 million people and not endowed with any natural resource (CIA, 2018). This lack of natural resources moulded a dynamic model of productivity and efficiency led by the service industry (Ministry of Foreign Affairs Singapore, 2018).

Chye (2014) shares that Singapore's Liveability Framework is centred on three main outcomes, namely, competitive economy, sustainable environment and high quality of life. This is represented in Fig. 3.1. The three outcomes form the basis for national policies formulated, with a specific focus on their completion. The success of these outcomes is tied in two systems of planning: (1) integrated master planning and development and (2) dynamic urban governance.

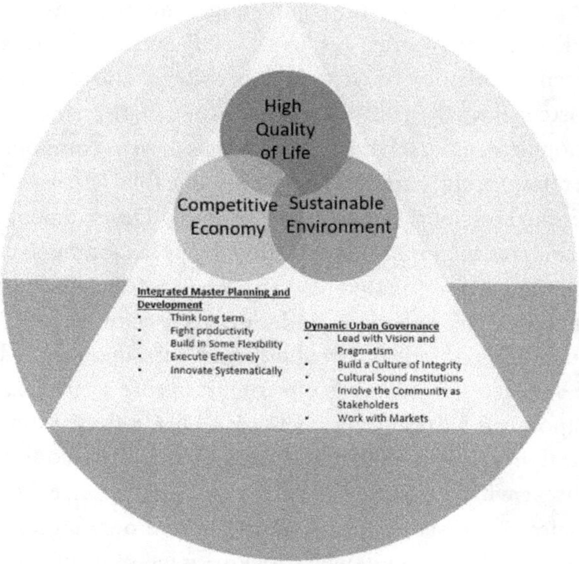

Fig. 3.1: The Singapore's Liveability Framework.

THE INTEGRATED MASTER PLANNING AND DEVELOPMENT

The Integrated Master Plan, led by the Urban Redevelopment Authority (URA), is structured such that it allows for long-term thinking, in an integrated, balanced and dynamic environment that promotes innovation and flexibility. It was first conceived in 1971 under the name Concept plan and was to guide the country's development agenda for the next 50 years. Under this were detailed different aspects of how the available land and resources were to be utilised for upcoming decades. The plan stipulates statutory land-use zoning plan, allowable use and density, hence facilitating transparency and fostering unfettered development. Its driving force, as noted by Tng and Tan (2011), is a far-sighted, holistic and comprehensive planning strategy that is geared towards the aforementioned three outcomes. Through public consultation, where citizens and planning stakeholders exchange ideas on how urban development needs to be undertaken, the master plan is reviewed every five years and is streamlined to accommodate new changes in the wake of upcoming new challenges. The review, as noted by Chye (2014), is meant to ensure that the country remains consistent in its endeavour to maintain a sound balance between density and liveability. This engages in an inter-agency relationship where different government agencies pursue individual goals but at the same time collaborate to ensure working towards a holistic plan.

The key areas of focus in the master plan include housing, community, transport, economy, recreation and identity, which are set to create a liveable environment. With this, the physical and social amenities and services are made available to residents of different parts of the country. Singapore has been known to maximise the compact urban form by adopting a 'housing for all' policy, where over 90% of the population own their own homes in high-rise apartments that are well planned and built. Within residential areas, social amenities are available with an increased pollution and congestion mitigation. Similarly, this strategy allows for substantial greener spaces that are ubiquitous in the country (UN Environment, 2018). In addition, it allows for a diversity of use and income where neighbourhoods are characterised of mixed-income groups and cultures (Chia, Li, & Yang, 2017), hence promoting an inclusive environment with job creation, opportunities, thriving businesses and provision of leisure.

DYNAMIC URBAN GOVERNANCE

The inclusion of 'dynamic urban governance' in the liveability framework as the second principle facilitates the realisation of the following three desirable outcomes: competitive economy, sustainable environment and high quality of life. While the political leadership model is contested, it is believed to achieve liveability objectives, guided by vision and pragmatism, integrity, commitment to sound institution, community participation and open to working with the markets (Civil Service College & Centre for Liveable Cities, 2014). To date, these tenets have been observed piously as the political leadership has shown consistency in vision and commitment to making Singapore an attractive, liveable city (UN Environment, 2018), and the desired outcomes are testament to this (Heritage, 2018; OECD, 2011; UNCTAD, 2018). An example of this is captured by Ganguli and Ebrahim (2017) who demonstrated that Singapore has managed to develop a medical model that promotes medical tourism, complementing its lack of natural resource and ultimately promoting economic diversification. Other supports that the status of Singapore moved from being that of a squalid settlement to that classified as modern, well lit, connected, greener and even further: a tourist destination city (S. C. Teo, 2014). The city is now adopting the smart city concept with advanced application of Artificial Intelligence (AI), Robotics and Big Data technologies (Huiling & Goh, 2017).

The strong leadership model ensures the continuity of large, complex and significantly challenging projects of national importance, such as the relocation of industries into areas further from residential areas, institution of laws, reducing congestion and pollution mitigation (UN Environment, 2016). Similar efforts have been made through the NEWater project that enables reclamation of used water into high-grade, ultra-clean water for consumption (Lee & Tan, 2016).

Jon Quah (2017) shares that integrity virtues have been observed especially regarding combating corruption. He attributes this to the political goodwill shown by Singaporean leaders and to the commitment of their anti-corruption agency. Institutions in Singapore are known to be corruption free with virtues of honesty and commitment (Ministry of Foreign Affairs Singapore, 2018). There is a strict application of law, but importantly, the majority of public leaders have demonstrated rationality in decision-making and national interests have superseded individual gratification (Yew, 2015b). One area that has uplifted the Singapore's Liveability Framework is the realisation of the contribution of the community, and its forging a future plan for the city. Citizens are seen as consumers of governmental projects and their involving in the early stages fosters behavioural change and early acceptance. This model of public engagement also contributed to making Singaporeans more environmentally conscious (Shen, Zhang, & Zhao, 2017).

URBAN METABOLISM AND SUSTAINABILITY

The concept of urban metabolism gained traction in recent decades due to its impacts on the urban planning, as is evident in the case of Singapore. Coined in 1965 by Wolman (1965), urban metabolism dwells in the question of how cities operate and explores the analogy of the internal processes of biological organisms which allow them to operate, grow and reproduce. He argued that cities require regular flow of resources such as water, materials, energy, people and food into the city and at the same rate a regular outflow of wastes. Cuddihy, Kennedy, and Engel-Yan (2008) later defined it as the sum of technical and socio-economic processes that occur in cities, resulting in growth, production of energy, and elimination of waste. It is also understood as the 'the assessment of the amount of resources produced and consumed by urban ecosystems' (Conke & Ferreira, 2015), as considered as a framework for modelling diverse urban fabrics in relation to sustainability and human activities (Allam & Newman, 2018b; Li & Kwan, 2018; Restrepo & Morales-Pinzrame2018). This concept is grounded on understanding and expounding the agenda of sustainable cities.

Singapore underlines the use of this concept and couples it with technology to help city managers and other stakeholders understand how different resources within the city fabrics are exchanged and how outputs of the urban fabric can be incorporated in other sectors as an input. Similarly, results from this analysis form an integral basis for policy formulation in regard to how resources are to be optimally used so as to maintain a safe balance between city growth, human activities and environmental integrity.

Enforcing a technological dimension in urban metabolism has ensured a better management of the recycling of waste and interconnection of flows across complex urban systems. This in line provides for increased sustainability levels to be achievable. The advantages that modern city planners and managers have are the availability of real-time data yielded from the interconnections of urban fabrics via technologies such as AI and Big Data, hence ample information for informed sustainability policies (Choong, 2018). However, while the advent of technology has been widely adopted in current times, the sustainability dimension has always been at the core of Singapore's restructuring drive.

Lee Kuan Yew was keen to ensure the national growth of economy, society and politics, and, at the same time, maintain a healthy and sustainable environment. Today, Singapore is known as 'A City in a Garden' (Chye, 2014). Attaining this was a daunting task since most of the resources required to sustain a sustainable city were initially missing. Access to clean water, nutrients, construction materials and other basic resources was missing. This prompted ambitious projects of national scale like the rehabilitation of the heavily polluted Singapore River.

Today the Singapore River is waste free with clean water and even hosts aquatic life (Joshi, Tortajada, & Biswas, 2012), as illustrated in Fig. 3.2, and the riverfront is well designed with promenades as illustrated in Fig. 3.3.

The project also entailed changing the residential set-up of high-rise buildings set to welcome a booming population, which has now reached over five million people. This led to securing of numerous patches of green public spaces that were turned into parks and recreation centres. In addition, the government achieved the objective of reduced and efficient water and energy consumption. Singapore relied on imported water from Malaysia but with its NEWater project and

Source: Image by Fahrul Azmi.

Fig. 3.2: The Singapore River Today.

Source: Image by Taylor Simpson.

Fig. 3.3: River Promenade.

concerted effort to harvest and store rainwater, it has managed to secure the city's water demand. In addition, reclamation of the rivers in the city has ensured that the city has sufficient water for irrigation, hence contributing significantly in ensuring continued supply of required nutrients in the city. The available water is optimally utilised with buildings, both residential and industrial and farms all adopting greener technologies. After use, the polluted water from these sources is recycled and reused (S. C. Teo, 2014). This cyclical flow ensures sustainable use of water and maintenance of the environment.

With over 90% of the population owning homes and booming industrial development, Singapore's energy consumption is increasing. The risk associated with this is that the country has for long relied on natural gas as its main source of energy. Concerted efforts have been made to ease the country's burden of non-renewable energy to greener sources and renewable gas. One major step towards this has been to ensure that buildings adopt the Building and Construction Authority's (BCA, Singapore) Green Mark certification, a scheme introduced in 2005. This certification ensures that measures are enforced to reduce water and energy usage and in improving indoor environments for a healthy workplace (BCA Singapore, 2018). Greener technologies, including the use of sensors, allow for real-time energy monitoring, installation and use of devices and equipment that are energy efficient. The adoption of alternative energy production like solar and wind energy and using environmentally friendly construction materials are also in place. In regard to solar, Singapore is yielding over 135 megawatt-peek (MWp) and is projected to reach 350 MWp by 2020 (Hean, 2017). As the climatic conditions of Singapore are favourable for the installation of alternative greener and cleaner sources of energy, industries and institutions are encouraged to adopt a more sustainable transition of energy. To ensure that the impacts of fossil fuels are reduced, motor vehicle taxes were introduced to encourage the use of public means of transportation, thus reducing automobile dependence and its unhealthy reliance on fossil fuel. N.-C. Chia and Phang (2001) contend that this approach increased the adoption rate of bicycle usage in disfavour of private vehicles, hence leading to smoother traffic flows, reduced pollution, health living and ultimately improved sustainability and economic growth.

Due to an increasing population, industrialisation, urbanisation and increased disposable income, the amount of waste generated in Singapore has increased exponentially. For instance, Bai and Sutanto (2002) claim that between 1972 and 2000, solid waste increased from 0.74 million tonnes to 2.8 million tonnes. Albeit this reality, models of waste management were implemented and have proved successful. The reduce, reuse and recycle model is enforced where even ashes from incinerated waste are used. In addition, the Government provides incentives to industries to reduce waste generation through the adoption of green technologies. The generated waste is collected and incinerated, and the collected ashes are disposed at the Semakau landfill, which is basically an island off the coast of Singapore, and the disposal of ashes there contributes to the infill of land for future development (Chan, 2016). The Semakau Island, promoted as the world's first offshore landfill, is featured in Fig. 3.4.

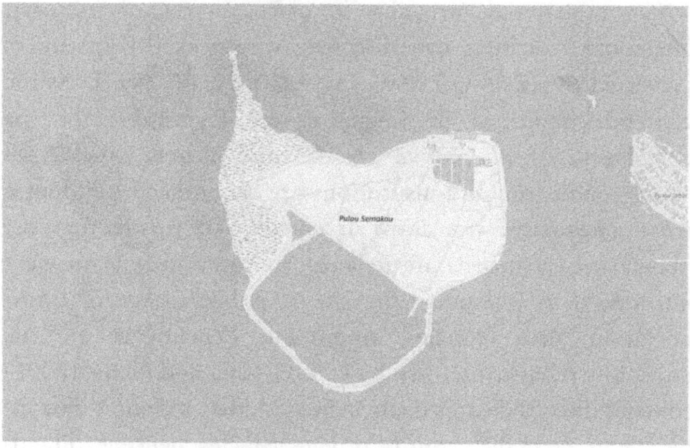

Source: © OpenStreetMap contributors.

Fig. 3.4: The Semakau Island.

There is thus a spirited effort to ensure recycling practices are emphasised to reach 70% nationally as advised by R. B. H. Tan and Khoo (2006). The requirement of Government is that all residents and industry report data on waste generated to better plan and manage waste collection and for a better enforcement of sustainable strategies to ensure the usage of waste by-products, like in the case of the Semakau Island.

TECHNOLOGY

The use of technology in the management of cities has had positive outcomes. Technologies such as Big Data, AI, Robotics, Machine Learning, Internet of Things (IoT), Blockchain etc. have greatly impacted traffic monitoring and control, healthcare improvement, service delivery, land management and control and safety and security improvement (Allam et al., 2019; Barns et al., 2016; PWC, 2018). Those have been noted to have a considerable impact on most spheres of the urban fabric, including the economy, social life, politics, legal framework, the environment and technology itself. Singapore has made us of advanced technology in the transformational stage from a third world country characterised of meagre resources to a modern, developing city-state that serves as a benchmark even for developed countries (S. Teo, 2014).

Amongst the transformations visible in Singapore are the digitisation and connectivity that are evident both in virtual and physical spaces. These are tied to its liveability framework that uses the city to thrive in a competitive economy, assuring its citizens a high quality of life in an environment that is clean, safe and sustainable (Civil Service College & Centre for Liveable Cities, 2014). The digitisation process is encapsulated in the e-Governement Action Plan (Government of Singapore,

2000) that aims at developing the country into a 'Smart nation', as stated in the Digital Government Blueprint (DGB) (GovTech Singapore, 2018). In addition, the digitisation strategy helps the country to take advantage of new technologies to address labour force constraints in an environment challenged by an ageing workforce. The strategy is also geared towards transforming different government sectors, hence achieving the core objectives captured in the liveability framework of the country. Ha and Coghill (2008) share that the service delivery in Singapore is characterised of accessibility, stakeholder-centricity, rapidity, efficiency, relevancy, digitally enabled, secure and cost-effectiveness courtesy of digitisation and government commitment. Unlike other countries where ICT integration is championed by the private sector (GovTech Singapore, 2018), in Singapore this is steered by the government. To ensure its success, the government has focused on three key initiatives including Smart Nation Programme, e-Government Action Plan (Government of Singapore, 2000) and the Infocomm Media 2025 as documented by EDB Singapore (2017).

One key area that has transformed as a result of digitisation is the public service (GovTech Singapore, 2018) which forms a key component of the government in its bid to achieve its developmental vision. It is supported in the DGB (GovTech Singapore, 2018) that the public service is facing challenges of workforce constraint, ageing and naivety in the use of ICT amongst some workers. The digitisation of this sector is further made to ensure a highly trained and adaptive workforce. Similarly, it provides citizens and businesses with an increase in the quality of services that are seamless and secure, hence building public trust, confidence and support.

Another area that has seen considerable improvement in the application of ICT is land management and monitoring (Chye, 2014). This is due to the interest of government in land management due to the limited size of the country against the increasing demand the real estate sector experiences from the surging population and the need for environmental sustainability. Land is a precious commodity in Singapore, hence the need to optimise it for economic, social and environmental benefits (Yew, 2015b). Amongst services digitised are land registration, titles registration systems, housing registration systems, Torrens systems, survey services and street directory (Ho, 2009; Singapore Land Authority (SLA), 2018). By digitising the land sector, the government has managed to improve efficiency monitoring and control land use especially in regard to residential buildings. Data from different quarters such as the CLC, BCA, Land Data Hub, Singapore Land Authority (SLA) and the private sector have allowed the government to track performance in demand and supply, which has helped to increase home ownership. Subsequently, the government has also managed to enforce the adoption of green technologies where over 80% of the buildings have been approved by BCA to have complied with requirements to reduce water and energy consumption (BCA Singapore, 2018). The use of data is also made in the prioritisation of water and energy distribution and in the monitoring of waste management processes.

SAFETY

The safety levels of Singapore have contributed to its brand of liveable city and as a touristic destination of choice. According to a police news release summarising the annual crime data for the year 2016, the overall crime rate had decreased by over 2.6% (Singapore Police Force, 2016). In 2017, the same agency reported a further 1% decrease. Crimes common in city centres such as robbery, housebreaking, pickpocketing and motor vehicle-related thefts have also been on the decrease (Singapore Police Force, 2017). These reports are affirmed by Economist Intelligence Unit (The Economist Intelligence Unit, 2017), which ranks Singapore in second position behind Tokyo in respect to Safety in Asian countries. The high score of 89.64% is noted to being consistent since 2015. In terms of law and order, Gallup World Polls ranked Singapore at the top in 2018 for the third year with a high of 97%. The author also report that 94% of Singaporeans interviewed said they were comfortable walking without company at night, without any fear, thus showing that its citizens are conscious of the safety levels of their city (Gallup, 2018).

The position of a safe haven given to Singapore is not incidental as explained in the police news release (Singapore Police Force, 2016, 2017). Security agencies and other stakeholders have capitalised on technologies such as Big Data and AI to ensure its citizens, business, and boundaries are safe, especially from the threat of cybersecurity, terrorism, organised crimes and traffic, amongst others. Smart devices such as cameras, sensors, anomaly alert and monitoring devices for maritime security and other monitoring devices have been strategically connected in various places, like in residential area, roads, airports and the port to monitor and send real-time data to the centralised command centres, hence quick responses (Leslie, Adams, Chin, & Tse, 2017; Singapore Police Force, 2016). Also, due to the good working relationship between the security agencies, private sector and the general public, reliable data and photo are sent to the security forces depicting areas where security or safety may be at risk (Ministry of Home Affairs Singapore, 2018). Security forces take advantage of digitisation in Singapore to share educative videos and posts via different platforms. According to the police news release in 2016, more than 1.3 million people have viewed the posts (Singapore Police Force, 2016). Availability of these monitoring devices, the collaboration with general public and pro-active stance of the forces and stringent security policies like the Cybersecurity Act 2018 (Ter, 2018) contribute to the safety index rating of the country.

RESILIENCE

Besides the contribution of cities to a country's economic, social, technological and political dimensions, they have impacted significantly to the environment. In the debate of urban metabolism, the clarion call is a sound balance between growth and sustainability. The inflow of resources and consequent outflow of waste materials need to be optimal and in favour of quality of life for the citizens characterised of high economic growth, stability and sustainable environment. Allam and Dhunny (2019) opine that the choices made in a bid to make cities more attractive, smart and liveable are key to addressing climate change both at a local and global sphere, and these add the benefits of having resilient and adaptable urban environment. However, the endeavour to have quality and modern urban infrastructure sometimes contributes in accelerating climate change consequences just as much as they can help in mitigating and delaying climate change (OECD, 2014).

In most case than not, actions taken in the cities are skewed towards accelerating the threat of climate change. Most cities are home to industries, most of the automobiles and high population, contributing significant amount of effluents that are released in water bodies, air and on land. Most of these wastes lead to an increase of greenhouse gases (GHGs) that impact the ozone layer, leading to climate change (Hoornweg, Sugar, & Trejos-Gomez, 2011). The demand for energy to power homes, businesses and urban life is another factor that makes cities forerunners in the climate change debate (Dhunny, Allam, Lobine, & Lollchund, 2019). Though there is evident of concerted effort and agreement from leaders drawn from all spheres and parts of the world to reduce emissions to manageable levels, most cities still heavily rely on fossil fuel for their energy production (Cottrell, Fortier, & Schlegelmilch, 2015; Covert, Greenstone, & Knittel, 2016) and some cities argue that it is crucial in maintaining some vital economic activities like employment opportunities (Covert et al., 2016).

The World Bank (2010) supports that the impact of climate change on cities will manifest on permanent and immobile structures such as bridges, subways, transportation infrastructures and buildings, amongst others, making cities more vulnerable. Similarly, the high population in cities is at risk due to severe threats of reduction in food supply, increasing temperatures, decreasing water supply, flooding and fogs, thus impacting on urban structures, systems and residents (Hallegatte, Green, Nicholls, & Corfee-Morlot, 2013; Li, Yang, Shi, & Pu, 2012; OECD, 2014).

With the knowledge of how the impacts of climate change can affect the cities, there have been efforts to mitigate and reduce these impacts. For instance, a study by Kennedy, Demoullin, and Mohareb (2012) on six major cities (Toronto, London, New York, Berlin, Boston and Seattle) established that there was an average reduction in GHG emissions of 0.27 t CO_2e/capita per year in these cities, which helped in increasing the liveability of those cities and in turn generate higher economic throughput. Gaignh helped in increasin2012), shared how the concept of compact cities is being implemented in major cities to reduce the environmental impacts caused by the transport system and urban sprawl. The approach could be increased in efficiency with the adoption of measures such as use of energy efficiency devices, use of alternative transport and green technologies. Singapore being a coastal city (Ziegler, Chan, Chuah, Dabrowski, & Varis, 2018), and with fewer natural resources, has taken spirited efforts to make sure its citizen, or infrastructural, food and other primary systems, is not affected by climate change, thus increasing its resilience. Its residential planning has adhered to compact buildings characterised of high-rise that houses substantial household and construction methods and materials that are environmentally friendly (Ong, 2017). The buildings are integrated with systems that ensure optimal water and energy usage and smart waste management. Besides saving land for environmental purposes and to accommodate the rising population, this building strategy has allowed Singapore to contain urban sprawl which has a far-reaching impact on climate change (Newman, 2010; Ong, 2017; Ziegler et al., 2018).

Water scarcity in Singapore has been an issue of concern and it was noted that the impacts of climate change could further exacerbate dire consequences (Ministry of Foreign Affairs Singapore, 2018). To ensure the city remains resilient in terms of water supply along with mitigating the impacts from climate change, Singapore has been taking pro-active steps. First, the cleaning of its rivers ensures that sufficient water is available for farming and maintaining the green ecosystem of the city (S. C. Teo, 2014). A careful water management provides water for consumption and feeds numerous other components of the urban system. Further to this, disposed water from residential areas and other sources is recycled through the NEWater project (Ministry of Foreign Affairs Singapore, 2018). The water from the Singapore River, contained by a mechanical dam as illustrated in Fig. 3.5, is contained and recycled. This is further made possible due to the integration between water supply, sewer lines and

Source: Image by Shaah Shahidh.

Fig. 3.5: The Singapore River Dam known as the 'Marina Barrage'.

drainage system manned by Singapore's national water authority (Civil Service College & Centre for Liveable Cities, 2014). Following the upgrading of the river, sites on river edges have been seen as extremely desirable for high-rise development as shown in Fig. 3.6.

Further to this, the government has ensured an extensive plan for maximising rainwater harvesting (PUB, 2018) and has also been engaging in desalination projects to reclaim seawater for household consumption, for agricultural

Source: Image by Chuttersnap.

Fig. 3.6: The Skyline Development along the Singapore River.

activities and for construction purposes. Through technology, public and private stakeholders have managed to advance membrane technologies, hence reducing the cost of desalination which has enabled the addition of over 130 million imperial gallons of supplied water per day from its three desalination plants (PUB, 2018; S. C.; Teo, 2014).

Land and resource limitation, coupled with climate change, would render Singapore drought prone, especially due to shortage of water, flooding and other environment-related problems. These problems compounded with a growing population and emerging middle-class stresses on agriculture and food supplies. For this reason, as explained by the Singapore Prime Minister, Mr Lee Hsien Loong, the government has been pro-active in ensuring innovative and technologically backed farming methods are implemented (Ministry of Foreign Affairs Singapore, 2018). These are meant to complement the imports on food, of which Singapore heavily rely one (Tortajada & Hongzhou, 2016). With its own supply, the country can supplement nutrition, hence supporting the healthcare sector in its preventive efforts (Ludher, 2016). Similarly, having enough food from own source would caution the country against external food price shocks (Rut & Davies, 2018), rendering a more resilient economic landscape. Amongst the projects explored in this line include the Sino-Singapore Jilin Food Zone (SSJFZ), that is, a collaboration between the Governments of Singapore and China. This project, whose first products were received in Singapore in 2016, is aimed at ensuring high-value and integrated agribusiness Research & Development, animal and pasture input, production and processing. This project has been classified as part of the Singapore's food security strategy, including overseas farming aimed at compensating for the limited agricultural land in the country (Tortajada & Hongzhou, 2016). A vertical farming project, named Sky Green, also allows for the cultivation of fresh vegetables using the carefully managed amounts of water, space and energy (Civil Service College & Centre for Liveable Cities, 2014; Ministry of Foreign Affairs Singapore, 2018).

BIOPHILIA

The concept of biophilia was devised by Wilson (1984) who argued that human beings have an innate attraction to nature and living things. Newman (2014) shares this belief and defined it as the love for living systems. Others explain it as the purposeful integration of natural elements in urban landscapes (Beatley, 2011; el-Baghdadi & Desha, 2017). The application of this concept to cities is understood to have numerous benefits like the reduction in energy demand in buildings, improvement in air quality, supporting urban biodiversity, reduction of noise, improvement in aesthetic aspects of a building and in the management of urban storm water surge, and adding competitiveness for increasing revenues for property developers, amongst others (Kotzen, 2018; Newman, 2014).

Notwithstanding these benefits coupled with advanced architectural technologies, many cities are still 'concrete jungles', as former Prime Minister Lee Kuan Yew referred to them (Yew, 2015b). The reasons behind this are diverse and many. On the forefront is the demand for residential and office spaces that is warranted by an increasing urban population and middle class. Also, the need for infrastructure expansion to accommodate increasing traffic, power lines, water distribution channels, sewer line etc. Ngom, Gosselin, and Blais (2016) classified these factors into socio-economic, demographic and geographic processes and argue that they pose the greatest challenge to availability of green spaces in most cities. The lucrativeness of the real estate industry coupled with weak land-use and control policies has far-reaching consequences on availability of green spaces in cities. Pawell (2015) explains how city planners, developers and government planners in some cities value tax revenues accrued from developments, hence perceiving any undeveloped space as forfeited revenue. Consequently, Nicholls (2005) shares that those advocating against the availability of green spaces invoke the notion of privacy infringement of those who reside near greenways from strangers and burglars. Crompton (2001) and McNeur (2017) further argue that establishing green spaces like parks and open spaces are expensive expedition that do not contribute any economic return to the community but only results in increase in land rates, taxes and increased land value in the areas, hence discouraging development.

However, in Singapore, green spaces are cherished, valued and encouraged. Indeed, the vision of having a greener

Singapore was borne and engrained in the endeavour to create a 'garden city' and started with tree-planting initiative steered by the former Prime Minister Mr Lee Kuan Yew (Ministry of Foreign Affairs Singapore, 2018). The vision materialised despite the physical challenge of limited land, urbanisation and increasing population. Since then, the government has been in the forefront in supporting green initiatives including the award of incentives and subsidies to developers and groups adopting innovative strategies that promote the culture of greener spaces. There have been numerous initiatives to promote this trend. One of such includes the Singapore Green Plan 2012, which was advanced by the Ministry of the Environment and Water Resource (Ministry of the Environment and Water Resources, 2018). The core mandate was to conserve the environment and its resource, and reclaim unlawfully developed areas, gathering relevant data and ensuring more green spaces are created. The efforts were supported by sound planning in areas like building and construction sector where most of the houses are compactly built by the Housing and Development Board (HDB), thus ensuring that sustainability and liveability goals are strictly adhered to. To complement these efforts, the BCA adopted mandatory requirements for new buildings, renovations and office interiors (Civil Service College & Centre for Liveable Cities, 2014). The call was for the adherence of optimal water and energy usage, use of environmentally friendly construction materials and ensuring adequate green spaces. New buildings are expected to include plant life either on roofs, vertical gardens or verdant walls (Kolczak, 2017). Compliant buildings are awarded a Green Mark signifying their adherence with these standards. In addition subsidies are also provided for developers who adopt the Sky Rise Greening Initiative (SRGI) that entails, including plants on the roofs and walls (Newman, 2014). One iconic project under the SRGI is the Khoo Teck Puat Hospital that integrated rooftop gardens, a 'green court' and lush cascading walls leading to its roof. Other notable designs include the Gardens by the Bay, which compromises of a lush green canopy (Fig. 3.7) and indoor waterfalls (Fig. 3.8); the HDB housing projects (Fig. 3.9); the Changi Airport Terminal 3 (Fig. 3.10); the Newton Suite and the Republic Polytechnic, taunted as a campus in a park (Newman, 2014).

Amongst the many initiatives include the Streetscape Greenery Master Plan that emphasises on the need for a continuous tree canopy towering over major road transport

Source: Image by Paul Wetzel.

Fig. 3.7: Gardens by the Bay.

Source: Image by Will Truettner.

Fig. 3.8: Indoor Waterfall, Gardens by the Bay.

systems. The greening vision is furthered by communities such as housing estates (both public and private), education institutions, NGOs, religious groups, businesses and charities, amongst others (Ministry of Foreign Affairs Singapore, 2018). Newman (2014) shares that the 'Community in Bloom' initiative forged in 2005 saw over 480 projects across Singapore, manned by different communities advancing green space initiatives and these have since then increased to approximately 1300 community gardens across the country (National Parks, 2018).

Source: Image by Chuttersnap.

Fig. 3.9: Green Roof on HDB Housing.

Source: Image by Peerapon Chantharainthron.

Fig. 3.10: Changi International Airport, Terminal 3.

DISCUSSION

From an extensive review of literature, it is clear that much have been achieved in Singapore, and as envisioned in the Singapore's Liveability Framework, the three desirable outcomes have guided the realisation of large complex national projects. These have been made possible through a strong political leadership with clear vision and commitment.

The various efforts to encourage environmentalism have contributed to social inclusivity on various levels. The review of literature revealed that the government has also ensured that its citizens live in a safe, clean and green environment with affordable clean water and energy. Inclusivity is not complete without the access to opportunity to contribute to governance on different affairs. Even though there are disputes about the strong leadership model, it was seen that the government of Singapore runs an open government where public participation is valued and where citizens' recommendations are translated into policies. This has led to a good working relationship between public and private sectors and the general public. For instance, there is a cordial relationship between the public and the country's security agencies; hence, the public feels free to share information, which can help in strengthening the security approaches. This has allowed Singapore to earn the accolade from reputable, international organisation (Gallup, 2018; The Economist Intelligence Unit, 2017) as amongst the most safety countries in the world. The safety has also been expounded in regard to environmental sustainability where Singapore is amongst the greenest countries in the world (Kolczak, 2017). This has been achieved through the concerted effort that involves the government, private sector, communities and individuals.

There are outstanding strides that have been made for the provision of quality education to every Singaporean. With quality education, the country has managed to tap into a fast-paced economy and created jobs for most of its citizens into different sectors. This has led to an unemployment rate of only 1.8% (OECD, 2011), a figure which is surprisingly low. Even with changing trends in the job market where demand for technological workforce accentuates, the government is committed to ensuring that its citizens are well trained and can benefit from these technologies. This supported the vision of ensuring a figure of above 90% home ownership as the citizens were economically equipped to honour loans.

Having experienced problems such as corruption, extreme poverty and pollution in the past, the approach of Mr Lee Kuan Yew, the first Prime Minister, was commendable. The adopted model demanded high levels of integrity, pragmatism, social inclusion, respect of institutions and partnership. By piously adhering to these principles, Singapore has prospered and today it enjoys a competitive economy, sustainable environment and a high quality of life (Ministry of Foreign Affairs Singapore, 2018). These unwavering stands in matters relating to the environment, economic growth, security and social inclusion have been used as a model by many countries as a benchmark of best practice (GovTech Singapore, 2018; Ministry of Foreign Affairs Singapore, 2018). On the same, it is clear that the high liveability levels in the country have helped the country attract substantial FDI that added to the contribution positively to the economic growth of the country (OECD, 2011). The same has also made Singapore a tourist attraction with substantial revenue accrued from this sector.

CONCLUSION

This chapter explores, through a review of literature, the Singapore's Liveability Model as the primary catalyst to the transformational change of Singapore. The strong political leadership of the country is seen to have contributed significantly to achieving its primary objectives as there was continuity in action. Today Singapore ranks highly in terms of sustainability, liveability and societal levels. Even though they suffer from a land shortage and from impacts of climate change, the resilience of the city-state has been greatly improved from the careful design of specific strategies oriented towards water, environment, food and others. This has greatly to do with its deep commitments to the concept of environmentalism and societal equity.

CONCLUSION

In these modern days, where cities are confronted by the challenges of increasing population, urbanisation and climate change, among others, there are visible increased global efforts to fashion cities and urban areas in a way that they can manage to achieve resilience. Nevertheless, despite having the advantage of advanced technologies, some countries and cities, especially in developing economies, are seen to be struggling on this front. For this reason, many have turned to study and, where possible, mirror the policies and development strategies applied in places that have achieved significant outcomes in a short time span. Through this, Singapore stands as a unique case study, whereby the country has managed to turn its fortune around in only one generation, from being a third world to a first world country in less than 50 years. From this book, the developmental history of Singapore is worth emulating in any economy, and those trying to borrow from it are perfectly in order but will need to ensure that proper and sensible calibration is done to achieve societal and political integration. Singapore, like many developed and developing countries, had a difficult start, especially due to the over-exploitation from power imperial rulers and colonisers who worked for self-gratification or for the sake of their home empires. Despite being a vocal trading hub for those in Southeast Asia and other parts of the world, it still lingered in poverty, corruption, underdevelopment and other numerous social and economic problems. Luckily, the country had some of its locals who were determined to change its fate, by seeking its independence from the British colonial masters, while commanding respect from its peers who looked down on the country. These included LKY and others who, through their agitation, managed to have the country adopt a new constitution and, at 1959, gain independence to self-rule from the British who had overstayed their visit since 1819. After independence what followed was a stream of developmental changes that saw the country transform from a meagre to a giant economy that today can be made to compete with the world's best. This was achieved despite the lack of natural resources or any prior important infrastructural development.

From the historical background discussed in Chapter 1, Singapore has been shown to have only its geographical location as its main resource. Being an island, it lacked any form of natural resources like water, forests and rare earth that it could exploit to improve its economy. To make things worse, the port, its only notable source of hope, was controlled by the British, thus leaving the locals at the mercies of these colonialists. On their part, the British exploited the locals by forcing them to work in the ports, while paying them very little. This led to the proliferation of squalid conditions in crowded slums that had no tangible supply of basic amenities. On the same, the quality of water from rivers was extremely polluted, and thus Singaporeans had to cling to the British for survival. Similarly, this was accentuated when Singapore was joined with its neighbours through the umbrella of Federation of Malaysia, where it encountered challenges like tax exploitation, corruption and unfair distribution of resources.

But as noted in the manuscript and explained at length in the literature, the country was not short of sober minds who would successfully steer the country out of the quagmire and into limelight of success. Since independence, Lee Kuan Yew stood as one of those and was shown to be a leader with determination, passion, vision and strong leadership skills. To begin with, he shaped the Housing and Development Board (HDB) and the Economic Development Board (EDB), which were tasked with the responsibility of providing decent housing for the Singaporeans and formulate economic-oriented blueprints that would guide the country to greater economic successes respectively. As of now, the HDB as demonstrated in the literature has ensured over 90% of the citizens own houses in a serene, safe and environmentally clean environment. The EDB on its part has guided the country through the master plan and liveability framework to resiliency in terms of economic growth, social inclusivity and environmental sustainability.

But on the forefront in all these development successes was LKY, whose leadership style, though criticised by some, earned him immeasurable accolades, both locally and across the globe. From the literature, it has comprehensively been shown how his attention, passion, desire and efforts were all directed towards the betterment of the country and the improvement of the liveability standards of his fellow countrymen. To achieve these noble objectives, he adopted a governance style that had little time for distractors; thus, some regarded him as an authoritarian leader and others, especially those who did not understand or experience the challenges of Singapore, interpreted him as a tyrant dictator. But to those who interacted with him, they knew him as 'soft' authoritarian who was pragmatic, a team player, mentor, patriotic and a passionate visionary who left nothing to chance to achieve what he intended.

By adopting his leadership style, he was able to force things to happen in a record time. For instance, the cleaning of Singapore River and other bodies of water, though first seen as a daunting task, was only done within a period of 10 years, and the result stands to date. On his own admission in a series of books cited in the literature above, the country was still vulnerable from many things, and most of its competing neighbours would have relished to see it fail; thus, he had to do whatever it took to ensure they succeeded. And from the literature, it is seen that he favoured elites and professionals whom he believed would share in his vision and also work as hard as him to ensure the programs initiated by the government were achieved. This is clearly demonstrated when he argued that his son, who later became the prime minister back in 2004, was to earn his position, but it would not be given on a silver platter. He believed in merit and not entitlement, and he emphasised this to the Singaporean by urging them to work hard and make sacrifices for their country if they were to be successful.

Through his leadership, LKY was able to inspire the country to participate in a massive nation-building exercise, and in less than 50 years the fruits of their labour were evident worldwide. The approval and legitimisation of his leadership, as demonstrated in the literature, is the trust shown by Singaporeans by voting for him and his ruling party (PAP) into leadership for five consecutive terms. As a person, he had been into the prime minister position for five terms and further 16 years into ministerial positions. Similarly, even the critics who disliked his leadership style admit that, nevertheless, he was able to achieve unimaginable developmental results for the country.

It was also demonstrated that the autocratic leadership style that LKY adopted is not for everyone, especially if the leader is on a self-enriching mission. A sizeable number of such leaders have failed terribly and left their country in desperate situations. But those who adopt this style to mimic LKY need to understand the underlying strategies that he, though authoritarian as some argue, was able to implement. Despite popular beliefs often associated with this governing style, it was noted that he cherished participatory decision-making, and none of the projects he initiated were devoid of public intervention. In particular, as shown in the discussion, he valued the engagement of professionals, civic groups, community organisations and elected leaders, both in the planning and execution stages of projects. Furthermore, the people he led liked his leadership style and the unique participatory strategies that the country eagerly implemented. For this reason, the public participated in projects and collectively those were refined to illustrate their wishes. For this reason, the neighbourhoods are said to be socially inclusive despite the multiplicity of cultures.

LKY's strong leadership and the development environment and aura that he created in Singapore also allowed room for innovative and bold policies that helped the country move from a third to first world country in just one generation. Such achievements are rare to find, for no other country has been able to attain the same, especially in a single generation; hence, Singapore remains the best example to deserve further studies. This has marvelled policymakers worldwide, and most leaders from different countries and cities have been keen to learn, understand and copy the same development strategies. But, from the literature, the achievement of Singapore is not only based on leadership style, but there are other factors like public participation, vision, commitment and sacrifice that need to be embraced.

In the twenty-first century especially, for those countries and cities that are trying to find their footing in the development paths, they, unlike Singapore had in the past, have technology as an added tool that they can exploit, especially to fast-track public participation and enhance developmental projects. Nevertheless, such tools will only work if such economies and cities are led by leaders who are flexible and visionary to drive long-term visions that are devoid of selfish interests. Such leaders need to be firm on their decision, and though they need to allow for freedom of speech, association and assembly, such should not be allowed to derail government development projects. Like in the case of Singapore, projects and programs should be steered by professionals and experts who have the prerequisite experience and passion to oversee their successful completion. Such, however, should be encouraged to take into account the societal needs and aspiration, such that all the programs have the blessing of the society, hence would be universally accepted and supported. By ensuring the society embraces the proposed developmental projects, such economies would experience rapid growth, hence achieve in promoting liveability, resilience and sustainability.

REFERENCES

AFJ Editor. (2010). China, Singapore food Zone inaugurates in Jilin. *Asia Food Journal*. https://www.asiafoodjournal.com/2010/05/china-singapore-food-zone-inaugurates-in-jilin/. Accessed on June 26, 2019.

Akingba, I. O. I., Kaliappan, S. R., & Hamzah, H. Z. (2018). Impact of health capital on economic growth in Singapore: An ARDL approach to cointegration. *International Journal of Social Economics*, *45*(2), 340–356. doi:10.1108/IJSE-12-2016-0376

Alam, N. (2015, March 23). How Lee Kuan Yew transformed Singapore. https://www.rsis.edu.sg/rsis-publication/cms/co15073-economic-legacy-of-lee-kuan-yew-lessons-for-aspiring-countries/.

Allam, Z. (2012). Sustainable architecture: Utopia or feasible reality? *Journal of Biourbanism*, 2(1), 47–61.

Allam, Z. (2018). Contextualising the smart city for sustainability and inclusivity. *New Design Ideas*, 2(2), 124–127.

Allam, Z. (2019). The emergence of anti-privacy and control at the nexus between the concepts of safe city and smart city. *Smart Cities*, 2(1), 96–105.

Allam, Z. (2020a). Data as the new driving gears of urbanization. In Z. Allam (Ed.), *Cities and the digital revolution: Aligning technology and humanity* (pp. 1–29). Cham: Springer International Publishing.

Allam, Z. (2020b). Digital urban networks and social media. In Z. Allam (Ed.), *Cities and the digital revolution: Aligning technology and humanity* (pp. 61–83). Cham: Springer International Publishing.

Allam, Z. (2020c). On culture, technology and global cities. In Z. Allam (Ed.), *Cities and the digital revolution: Aligning technology and humanity* (pp. 107–124). Cham: Springer International Publishing.

Allam, Z. (2020d). Privatization and privacy in the digital city. In Z. Allam (Ed.), *Cities and the digital revolution: Aligning technology and humanity* (pp. 85–106). Cham: Springer International Publishing.

Allam, Z. (2020e). Theology, sustainability and big data. In Z. Allam (Ed.), *Theology and urban sustainability* (pp. 53–67). Cham: Springer International Publishing.

Allam, Z. (2020f). Urban chaos and the AI Messiah. In Z. Allam (Ed.), *Cities and the digital revolution: Aligning technology and humanity* (pp. 31–60). Cham: Springer International Publishing.

Allam, Z. (2020g). *Cities and the digital revolution: Aligning technology and humanity*. Cham: Springer International Publishing.

Allam, Z., & Dhunny, Z. A. (2019). On big data, artificial intelligence and smart cities. *Cities*, *89*, 80–91.

Allam, Z., Dhunny, A., Siew, G., & Jones, D. (2018). Towards smart urban regeneration: Findings of an urban footprint survey in Port Louis, Mauritius. *Smart Cities*, *1*(1), 121–133.

Allam, Z., & Newman, P. (2018a). Economically incentivising smart urban regeneration. Case study of Port Louis, Mauritius. *Smart Cities*, *1*(1), 53–74.

Allam, Z., & Newman, P. (2018b). Redefining the smart city: Culture, metabolism and governance. *Smart Cities*, *1*(1), 4–25. doi:10.3390/smartcities1010002

Allam, Z., Tegally, H., & Thondoo, M. (2019). Redefining the use of big data in urban health for increased liveability in smart cities. *Smart Cities*, 2(2), 259–268.

Andelman, D. A. (2018). The global move from democracy to autocracy. June 25. https://edition.cnn.com/2018/06/24/opinions/trump-end-to-democracy-opinion-andelman/index.html. Accessed on June 27, 2019.

Austin, I. P. (2009). Singapore in transition: Economic change an political consequences. *Journal of Asian Public Policy*, 2(3), 266–278. doi:10.1080/17516230903204745

Bai, R., & Sutanto, M. (2002). The practice and challenges of solid waste management in Singapore. *Waste Management*, *22*(5), 557–567.

Barns, S., Cosgrave, E., Acuto, M., & Mcneill, D. (2016). Digital infrastructures and urban governance. *Urban Policy and Research*, *35*, 20–31.

Baum, S., Yigitcanlar, T., Mahizhnan, A., & Andiappan, N. (2017). Singapore government online: A consideration of E-government outcomes. *Journal of E-Government*, 3(4), 65–84. doi:10.1300/J399v03n04_04

BBC News. (2018). Singapore profile – Timeline. May 10. https://www.bbc.com/news/world-asia-15971013. Accessed on June 21, 2019.

BBC News. (2019). Sudan coup: Why Omar al-Bashir was overthrown. April 15. https://www.bbc.com/news/world-africa-47852496. Accessed on June 21, 2019.

BCA Singapore. (2018). About BCA Green Mark scheme. October 3. https://www.bca.gov.sg/greenmark/green_mark_buildings.html.

Beatley, T. (2011). *Biophilic cities: Integrating nature into urban design and planning*. Washington, DC: Island Press.

Berge, S. (2004). Lee Kuan Yew's son takes over as Singapore leader. August 11. https://www.telegraph.co.uk/news/worldnews/asia/singapore/1469191/Lee-Kuan-Yews-son-takes-over-as-Singapore-leader.html. Accessed on June 21, 2019.

BOI. (2015). *Smart city scheme guidelines*. Port Louis. edbmauritius.org/media/1674/smart-city-scheme-guidelines.pdf. Accessed on November 3, 2018.

Borschberg, P. (2010). *The Singapore and Melaka Straits: Violence, security and diplomacy in the 17th century*. Singapore: NUS Press.

Bose, R. (2005). *The end of the war: Singapore's liberation and the aftermath of the Second World War*. Singapore: Marshall Cavendish Editions.

Buckley, C. B. (1984). *An anecdotal history of old times in Singapore 1819–1867*. Singapore: Oxford University Press.

Burton, J., Montagnon, P., Brown, K., & Grant, J. (2015). Lee Kuan Yew, Singapore's founding father, 1923–2015. March 22.

https://www.ft.com/content/24145cfe-b89d-11e2-869f-00144feabdc0. Accessed on June 21, 2019.

Cangi, E. C. (1993). Civilizing the people of Southeast Asia: Sir Stamford Raffles' town plan for Singapore, 1819–23. *Planning Perspectives*, *8*(2), 166–187. doi:10.1080/02665439308725769

CEIC. (2019). Singapore Foreign Direct Investment. https://www.ceicdata.com/en/indicator/singapore/foreign-direct-investment.

Chan, D. (2018). What does a highly liveable Singapore mean? https://www.straitstimes.com/opinion/what-does-a-highly-liveablesingapore-mean. Accessed on June 23, 2019.

Chan, J. K. H. (2016). The ethics of working with wicked urban waste problems: The case of Singapore's Semakau Landfill. *Landscape and Urban Planning*, *154*, 123–131.

Chappell, B. (2017). Zimbabwe's Mugabe out of power for first time since 1980s as military takes control. November 15. https://www.npr.org/sections/thetwo-way/2017/11/15/564320495/zimbabwes-mugabe-is-out-of-power-for-first-time-since-1980s-military-denies-coup. Accessed on June 28, 2019.

Chen, K. (2016). Lee Kuan Yew and Singapore. In Y. C. Ning, W. Gungwu, & Y. Ying-shih (Eds.), *Lee Kuan Yew through the eyes of Chinese scholars* (pp. 35–45). Singapore: World Scientific Publishing.

Chia, N.-C., & Phang, S.-Y. (2001). Motor vehicle taxes as an environmental management instrument: The case of Singapore. *Environmental Economics and Policy Studies*, *4*(2), 67–93.

Chia, W.-M., Li, M., & Yang, T. (2017). Public and private housing markets dynamics in Singapore: The role of fundamentals. *Journal of Housing Economics*, *36*, 44–61.

Ching, L. (2015). A quantitative investigation of narratives: Recycled drinking water. *Water Policy*, *17*(5), 831–847. doi: 10.2166/wp.2015.125

Choi, F. (2018). How to establish a good government? Lessons from Lee Kuan Yew in Singapore and Deng Xiaoping in China. *Public Administration and Policy*, *21*(2), 107–119.

Chong, A., & Chan, S. (2017). Militarizing civilians in Singapore: Preparing for 'crisis' within a calibrated nationalism. *The Pacific Review*, *30*(3), 365–384.

Choong, K. W. (2018). A clean & green city – Waste management in Singapore. Waste Management Solutions Forum: An Ecowaste Exhibition Event 2018. https://www.ecowaste.ae/__media/WMSF-Presentations/Kwok-Wai-Choong_A-clean-and-green-city-Waste-Management-in-Singapore.pdf.

Chye, K. T. (2014). The CLC liveability framework. In K. T. Chye, T. B. Kwan, P. Somasundram, V. Oliveiro, J. Yong, & L. Lee (Eds.), *Liveable sustainable cities: A framework*. Singapore: Centre for Liveable Cities/ Civil Service College Singapore.

CIA. (2018). World factbook: Singapore. October 23. https://www.cia.gov/library/publications/the-world-factbook/geos/sn.html. Accessed on October 29, 2018.

Civil Service College & Centre for Liveable Cities. (2014). *Liveable & sustainable cities: A framework*. Singapore: Centre for Liveable Cities/ Civil Service College.

CLC. (2018). Singapore liveability framework. https://www.clc.gov.sg/research-publications/framework. Accessed on October 30, 2018.

Clements-Croome, D., Marson, M., Yang, T., & Airaksinen, M. (2017). Planning and design scenarios for liveable cities. In M. A. Abraham (Ed.), *Encyclopedia of sustainable technologies* (pp. 81–97). Amsterdam: Elsevier.

Conke, L. S., & Ferreira, T. L. (2015). Urban metabolism: Measuring the city's contribution to sustainable development. *Environmental Pollution*, *202*, 146–152.

Cottrell, J., Fortier, F., & Schlegelmilch, K. (2015). Fossil fuel to renewable energy: Comparator study of subsidy reforms and energy transitions in African and Indian Ocean Island states. *United Nations Office for Sustainable Development, Incheon, Republic of Korea*, *1*, 1–45.

Country Economy. (2019a). Country comparison Hong Kong vs Singapore. https://countryeconomy.com/countries/compare/hong-kong/singapore?sc=XE34.

Country Economy. (2019b). Country comparison Japan vs Singapore. https://countryeconomy.com/countries/compare/japan/singapore?sc=XE15.

Covert, T., Greenstone, M., & Knittel, C. R. (2016, February). Will we ever stop using fossil fuels? http://ceepr.mit.edu/files/papers/2016-003.pdf.

Crinson, M. (2017). Singapore's moment: Critical regionalism, its colonial roots and profound aftermath. *The Journal of Architecture*, 22(4), 689–709. doi:10.1080/13602365.2016.1204071

Crompton, J. L. (2001). The impact of parks on property values: A review of the empirical evidence. *Journal of Leisure Research*, *33*(1), 1–31.

Cuddihy, J., Kennedy, C., & Engel-Yan, J. (2008). The changing metabolism of cities. *Journal of Industrial Ecology*, *11*(2), 43–59.

Demir, M. (2017). Democracy v.s autocracy – A conundrum for Turkey. September 18. https://www.vocaleurope.eu/democracy-v-s-autocracy-a-conundrum-for-turkey/. Accessed on June 21, 2019.

Department of Statistics Singapore. (2019). *Singapore in figures*. Singapore. https://www.singstat.gov.sg/-/media/files/publications/reference/sif2019.pdf. Accessed on November 3, 2018.

Dhunny, A. Z., Allam, Z., Lobine, D., & Lollchund, M. R. (2019). Sustainable renewable energy planning and wind farming optimization from a biodiversity perspective. *Energy*, *185*, 1282–1297.

Doner, R. F., Ritchie, B. K., & Slater, D. (2005). Systemic vulnerability and the origins of developmental states: Northeast and Southeast Asia in comparative perspective. *International Organization*, *56*, 321–361.

Donnithome, G. C. A. (2003). *Western enterprise in Indonesia and Malaysia*. Abingdon: Routledge.

EDB Singapore. (2017). Government leads digital disruption in Singapore. May 8. https://www.edb.gov.sg/en/news-and-resources/insights/innovation/government-leads-digital-disruption-in-singapore.html. Accessed on November 2, 2018.

el-Baghdadi, O., & Desha, C. (2017). Conceptualising a biophilic services model for urban areas. *Urban Forestry and Urban Greening*, *27*, 399–408.

En, S. M. (2017). The Big Read: Far from people's minds, but food security a looming issue. May 26. https://www.todayonline.com/singapore/big-read-far-peoples-minds-food-security-looming-issue. Accessed on November 2, 2018.

Environmental Performance Index. (2019). Singapore. https://epi.envirocenter.yale.edu/epi-country-report/SGP. Accessed on June 26, 2019.

Farrell, B. (2017). *The defence and fall of Singapore.* Singapore: Monsoon Books Private Limited.

Feraco, A., & Müller-Wittig, W. (2014). *Singapore policy-making processes: The impacts of ICT to enhance public participation and gather meaningful insights.* Hershey, PA: IGI Global.

Findlay, R., & O'Rourke, K. H. (2007). *Power and plenty: Trade, war, and world economy in the Second Millennium.* Princeton, NJ: Princeton University Press.

Fishkin, J. S. (2018). *Democracy when the people are thinking: Revitalizing our politics through public deliberation.* Oxford: Oxford University Press.

Gaigné, C., Riou, S., & Thisse, J.-F. (2012). Are compact cities environmentally friendly? *Journal of Urban Economics*, *72*, 123–136. doi:10.1016/j.jue.2012.04.001

Gallup. (2018). 2018 global law and order. *Gallup World Polls*, pp. 1–11. https://www.insightcrime.org/wp-content/uploads/2018/06/Gallup_Global_Law_And_Order_Report_2018.pdf.

Ganguli, S., & Ebrahim, A. H. (2017). A qualitative analysis of Singapore's medical tourism competitiveness. *Tourism Management Perspectives*, *21*, 74–84.

Giap, T. K., Thye, W. W., & Aw, G. (2014). A new approach to measuring the liveability of cities: The global liveable cities index. *World Review of Science, Technology and Sustainable Development*, *11*(2), 176–196.

Gilchrist, K. (2019). Here are the world's best airports for 2019. https://www.cnbc.com/2019/03/28/singapores-changi-wins-worlds-best-airport-award-beats-toyko-seoul.html. Accessed on June 25, 2019.

Gleason, N. W. (2018). Singapore's higher education systems in the era of the fourth industrial revolution: Preparing lifelong learners. In N. Gleason (Ed.), *Higher education in the era of the fourth revolution*. Singapore: Palgrave Macmillan.

Global is Asian. (2017). Singapore's success in urban planning: Learning from father of city planning Dr. Liu Thai Ker. August 30. https://lkyspp.nus.edu.sg/gia/article/singapore-s-success-in-urban-planning-learning-from-father-of-city-planning-dr-liu-thai-ker. Accessed on June 28, 2019.

Government of Singapore. (2000, September 14). The Singapore e-government action plan. http://unpan1.un.org/intradoc/groups/public/documents/apcity/unpan002999.pdf.

Government of Singapore. (2018, July 20). Land Transport. https://www.mot.gov.sg/about-mot/land-transport.

GovTech Singapore. (2018). *Digital government blueprint: A Singapore government that is digital to the core, and serves with heart*. Singapore. https://www.tech.gov.sg/files/

digital-transformation/dgb_summary_june2018.pdf. Accessed on June 26, 2019.

Guo, R. (2018). Citizen participation for inclusive outcomes: It takes a village to plan a city. *Urbans Solutions*, *12*, 54–63.

Guriev, S. M., & Treisman, D. (2015). How modern dictators survive: Cooptation, censorship, propaganda, and repression. *SSRN Electronic Journal*, 1–62. doi:10.2139/ssrn.2571905

Ha, H., & Coghill, K. (2008). E-government in Singapore a SWOT and PEST analysis. *Asia-Pacific Social Science Review*, *6*(2), 103–130.

Haackle, J. (2005). *ASEAN's diplomatic and security culture: Origins, development and prospects*. New York, NY: Routledge.

Hack, K., Margolin, J. L., & Delaye, K. (2010). *Singapore from Temasek to the 21st century: Reinventing the global city*. Honolulu, HI: University of Hawaii Press.

Hallegatte, S., Green, C., Nicholls, R. J., & Corfee-Morlot, J. (2013). Future flood losses in major coastal cities. *Nature Climate Change*, *3*, 802–806.

Han, H. (2017). Singapore, a Garden City: Authoritarian environmentalism in a development state. *The Journal of Environment & Development*, *26*(1), 3–24. doi: 10.1177/1070496516677365

Han, F. K., Fernandez, W., & Tan, S. (1998). *Lee Kuan Yew: The Man and his ideas*. Singapore: Times Edition.

Hean, T. C. (2017, June 5). DPM Teo Chee Hean at the Ecosperity conference. https://www.pmo.gov.sg/newsroom/dpm-teo-chee-hean-ecosperity-conference#_ftnref5.

Heng, C. K. (2016). *50 years of urban planning in Singapore*. Singapore: NUS.

Heritage. (2018). *2018 index of economic freedom*. Singapore. https://www.heritage.org/index/country/singapore.

Herrman, T., & Lewis, R. (2017). What is livability? Sustainable cities initiative (SCI). https://sci.uoregon.edu/sites/sci1.uoregon.edu/files/sub_1_-_what_is_livability_lit_review.pdf. Accessed on October 29, 2018.

Hibbing, J. R., & Theiss-Morse, E. (2002). *Stealth democracy: Americans' beliefs about how government should work*. New York, NY: Cambridge University Press.

Ho, R. (2009). *Singapore's land administration in the electronic age* (pp. 1–23). Singapore: Singapore Land Authority.

Ho, P. (2013). Model of governance: Big government or big people? In S. H. Kang (Ed.), *Singapore perspective 2012: Singapore inclusive: Bridging divides* (pp. 117–124). Singapore: World Scientific Publication Co.

Hoornweg, D., Sugar, L., & Trejos-Gomez, C. L. (2011). Cities and greenhouse gase emissions: Moving forward. *Environment and Urbanization*, *23*(1), 207–227.

Horton, R. (2018). Offline: The invisibility of inequality. *The Lancet*, *392*(10163), 2424. doi:10.1016/S0140-6736(18)33110-6

Howley, P., Scott, M., & Redmond, D. (2009). Sustainability versus liveability: An investigation of neighbourhood satisfaction. *Journal of Environmental Planning and Management*, *52*(6), 847–864.

Huiling, E., & Goh, B. (2017). AI, robotics and mobility as a Service: The case of Singapore. *Field Actions Science Reports*, *1*(17), 26–29, Special Issue.

Hunt, D. V. L., Makana, L. O., Jefferson, I., & Rogers, C. D. F. (2016). Liveable cities and urban underground space. *Tunnelling and Underground Space Technology*, *55*, 8–20.

Hyman, G. (2015). Lee Kuan Yew's Enigma: Authoritarian yet a kind of democrat. https://www.csis.org/analysis/lee-kuan-yew%E2%80%99s-enigma-authoritarian-yet-kind-democrat. Accessed on June 28, 2019.

IMD. (2019). IMD world competitiveness ranking 2019: Singapore topples United States as world's most competitive economy. Lausanne. https://www.prnewswire.com/in/news-releases/imd-worldcompetitiveness-ranking-singapore-topples-united-states-as-world-s-mostcompetitive-economy-859915181.html. Accessed on June 27, 2019.

Index Mundi. (2019). Singapore – GDP per capita. https://www.indexmundi.com/facts/singapore/gdp-per-capita. Accessed on June 25, 2019.

Irvin, R. A., & Stansbury, J. (2004). Citizen participation in decision making: Is it worth the effort? *Public Administration Review*, *64*(1), 55–65.

Jha, A. (2018, January 31). "But what about Singapore?" Lessons from the best public housing program in the world. http://blogs.worldbank.org/sustainablecities/what-about-singapore-lessons-best-public-housing-program-world.

Jones, D. S. (2016). Governance and meritocracy: A study of policy implementation in Singapore. In J. Quah (Ed.), *The role of the public bureaucracy in policy implementation in five*

ASEAN countries (pp. 297–269). Cambridge: Cambridge University Press.

Joshi, Y. K., Tortajada, C., & Biswas, A. K. (2012). Cleaning of the Singapore River and Kallang Basin in Singapore: Human and environmental dimensions. *Ambio*, *41*(7), 777–781. doi:10.1007/s13280-012-0279-0

Juma, C. (2013). Development: Learning from Singapore's Lee Kuan Yew. *Analysis & Opinion*, January 20. https://www.belfercenter.org/publication/development-learning-singapores-lee-kuan-yew-0. Accessed on June 26, 2019.

Kennedy, C., Demoullin, S., & Mohareb, E. (2012). Cities reducing their greenhouse gas emissions. *Energy Policy*, *49*, 774–777.

Khew, C. (2015). Current measures against littering in Singapore. https://www.straitstimes.com/singapore/environment/current-measures-against-littering-in-singapore. Accessed on June 28, 2019.

Kolczak, A. (2017, February 28). This city aims to be the world's greenest. https://www.nationalgeographic.com/environment/urban-expeditions/green-buildings/green-urban-landscape-cities-Singapore/.

Kolesnikov-Jessop, S. (2010). Singapore looks to China for food security. September 27. https://www.nytimes.com/2010/09/28/business/global/28iht-rbofsing.html. Accessed on June 26, 2019.

Kotzen, B. (2018). Green roofs social and aesthetic aspects. In G. Pérez & K. Perini (Eds.), *Nature based strategies for urban and building sustainability* (pp. 273–281). Oxford: Butterworth-Heinemann.

Kwang, H. F., Ibrahim, Z., Hoong, C. M., Lim, L., Low, I., Lin, R., & Chan, R. (2011). *Lee Kuan Yew: Hard truths to keep Singapore going*. Singapore: Straits Time Press.

Lahiri, T. (2017). Singapore's new car cap will keep over 1,500 cars from hitting its road in a year. October 24. https://qz.com/1110049/singapore-is-capping-cars-because-its-running-out-of-room-to-build-roads/. Accessed on June 26, 2019.

Lambert, T. (2019). A brief history of Singapore. *The Foundation of Singapore*. http://www.localhistories.org/singapore.html.

Lee, H., & Tan, T. P. (2016). Singapore's experience with reclaimed water: NEWater. *International Journal of Water Resources Development*, *32*(4), 611–621. doi:10.1080/07900627.2015.1120188

Lee, J. (2018). Understanding authoritarian resilience and countering autocracy promotion in Asia. *Asia Policy*, *13*(4), 99–122.

Lee, K. Y. (2019). Lee quoted in "Without Mr Lee Kuan Yew, 'there would be no Garden City'" in SM En. *Today Online*, August 11. https://www.todayonline.com/rememberinglky/without-mr-lee-kuan-yew-there-would-be-no-garden-city.Accessed on August 1, 2019.

Leong, H. K. (2000). Citizen participation and policy making in Singapore: Conditions and predicaments. *Asian Survey*, *40*(3), 436–455.

Leslie, L. C. W., Adams, K. T. C., Chin, K. S., & Tse, H. C. (2017). Enhancing maritime security through data analytics. *DSTA Horizons*, pp. 64–69.

Leyden, J., & Raffles, T. S. (1821). *Malay annals*. London: A. & R. Spottiswoode.

Li, C. (2018). Connecting to the world: Singapore as a Hub Port. https://www.csc.gov.sg/articles/connecting-to-the-world-singapore-as-a-hub-port.

Li, H., & Kwan, M.-P. (2018). Advancing analytical methods for urban metabolism studies. *Resources, Conservation and Recycling*, *132*, 239–245.

Li, Z., Yang, J., Shi, C., & Pu, M. (2012). Urbanisation effects on fog in China: Field research and modeling. *Pure and Applied Geophysics*, *169*(5–6), 927–939.

Lim, J., & Lee, T. (2016). *Singapore: Negotiating state and society, 1965–2015*. New York, NY: Routledge.

Liu, T. K. (2016). Planning & urbanisation in Singapore: A 50- year journey. In H. C. Kiang (Ed.), *50 years of urban planning in Singapore* (pp. 23–44). Singapore: World Scientific Publishing. doi:10.1142/9789814656474_0002

Lloyd, T. O. (1996). *The British Empire 1558–1995*. Oxford: Oxford University Press.

Long, J. (2001). Desecuritizing the water issue in Singapore-Malaysia relations. *Contemporary Southeast Asia*, *23*(2), 504–532.

Luan, I. O. B. (2010). Singapore water management policies and practices. *International Journal of Water Resources Development*, *26*(1), 65–80.

Ludher, E. K. (2016). Singapore's smart governance of food. In M. Deakin, D. Diamantini, & N. Borreli (Eds.), *The governance of city food systems: Case studies from around the world* (pp. 131–154). Milan: Fondazione Giangiacomo Feltrinelli.

Maritime Port Authority Singapore. (2018). International Maritime Centre 2030: Strategic review. Singapore. https://www.gov.sg/~/sgpcmedia/media_releases/mpa/press_release/p-20170922-2/attachment/imc%202030%20strategic%20report.pdf. Accessed on November 3, 2018.

Mauro, B. W. D. (2018). Building a cohesive society: The case of Singapore's housing policies (Policy Brief No. 128). https://www.cigionline.org/sites/default/files/documents/PB%20no.128web.pdf. Accessed on June 27, 2019.

McNeur, C. (2017). Parks, people, and property values: The changing role of green spaces in Antebellum Manhattan. *Journal of Planning History*, *16*(2), 98–111. doi:10.1177%2F1538513216657563

Meadows, D. (1988, May 5). Singapore leads the good life under a benevolent dictator. http://donellameadows.org/archives/singapore-leads-the-good-life-under-a-benevolent-dictator/.

Menon, R. (2015, August 5). Ravi Menon: An economic history of Singapore – 1965–2065. Paper presented at the Singapore Economic Review Conference, Singapore.

Ministry of Foreign Affairs Singapore. (2018). Towards a sustainable and resilient Singapore. Singapore. https://sustainabledevelopment.un.org/content/documents/19271Singapores_Voluntary_National_Review_Report.pdf. Accessed on November 3, 2018.

Ministry of Home Affairs Singapore. (2018, August 17). Innovating to keep Singapore safe and secure. https://www.mha.gov.sg/about-us/innovating-to-keep-singapore-safe-and-secure. Accessed on November 3, 2018.

Ministry of the Environment and Water Resources. (2018, August 6). Singapore green plan. https://www.mewr.gov.sg/grab-our-research/singapore-green-plan-2012. Accessed on June 26, 2019.

Ministry of Trade and Industry Singapore. (2018). Speech by Minister Chan Chun Sing at the 11th Singapore International Energy Week 2018. October 30. https://www.mti.gov.sg/Newsroom/Speeches/2018/10/Speech-by-Minister-Chan-Chun-Sing-at-the-11th-Singapore-International-Energy-Week-2018. Accessed on November 4, 2018.

Ministry of Transport. (2019). Road network. https://www.mot.gov.sg/about-mot/land-transport/motoring/road-network. Accessed on June 24, 2019.

Mody, A. (1997). *Infrastructure strategies in East Asia: The Untold story*. Washington, DC: The International Bank for Reconstruction and Development/The World Bank.

Morgenbesser, L. (2016). The autocratic mandate: Elections, legitimacy and regime stability in Singapore. *The Pacific Review*, *30*(2), 205–231.

National Parks. (2018). Community in bloom initiatives. https://www.nparks.gov.sg/gardening/community-in-bloom-initiative. Accessed on November 3, 2018.

Newman, P. (2010). Green urbanism and its application to Singapore. *Environment and Urbanization ASIA*, *1*(2), 149–170. doi:10.1177%2F097542531000100204

Newman, P. (2014). Biophilic urbanism: A case study on Singapore. *Australian Planner*, *51*(1), 47–65.

Ngom, R., Gosselin, P., & Blais, C. (2016). Reduction of disparities in access to green spaces: Their geographic insertion and recreation functions matter. *Applied Geography*, *66*, 35–51.

Nicholls, S. (2005). The impact of greenways on property values: Evidence from Austin, Texas. *Journal of Leisure Research*, *37*(3), 321–341.

Norouzian-Maleki, S., Bell, S., Hosseini, S., Faizi, M., & Bahram, S. (2018). A comparison of neighbourhood liveability as perceived by two groups of residents: Tehran, Iran and Tartu, Estonia. *Urban Forestry and Urban Greening*, *35*, 8–20.

OECD. (2011). Singapore: Rapid improvement followed by strong performance. In *Strong performers and successful reformers in education: Lessons from Pisa for the United States* (pp. 159–175). Paris: OECD.

OECD. (2014). *Cities and climate change: National governments enabling local action*. OECD policy Perspectives. Paris: OECD.

Ong, K. C. G. (2017). Sustainable construction for Singapore's urban infrastructure – Some research findings. *Procedia Engineering*, *171*, 14–21. doi:10.1016/j.proeng.2017.01.305

Ooi, G. L. (2005). *Sustainability and cities: Concepts and assessments*. Singapore: World Scientific. Accessed on June 26, 2019.

OpenStreetMap. https://www.openstreetmap.org/copyright.

Pawell. (2015). Literature survey: Green space and property values by Natasha Catrakilis. April 28. https://sites.duke.edu/urbaneconomics/?p=1441. Accessed on November 3, 2019.

PUB. (2018). Singapore water story. October 29. https://www.pub.gov.sg/watersupply/singaporewaterstory. Accessed on October 29, 2018.

PWC. (2018). Blockchain: The next innovation to make our cities smarter. FICCI. https://www.pwc.in/assets/pdfs/publications/2018/blockchain-the-next-innovation-to-make-our-cities-smarter.pdf.

Quah, J. (2007). *Combating corruption Singapore style: Lessons for other Asian countries*. Baltimore, MD: School of Law, University of Maryland.

Quah, J. (2017). Singapore's success in combating corruption: Lessons for policy makers. *Asian Education and Development Studies*, *6*(3), 263–274. doi:10.1108/AEDS-03-2017-0030

Quah, J. (2018). Why Singapore works: Five secrets of Singapore's success. *Public Administration and Policy, 21*(1), 1–18. doi:10.1108/PAP-06-2018-002

Ramdas, K. (2015). Beyond 'backwater' island to global city vision? Rethinking feminist geographies for Singapore after Lee Kuan Yew. *Geoforum*, *65*, 108–111. doi:10.1016/j.geoforum.2015.07.016

Rana, P. B., & Lee, C.-Y. (2015). Economic legacy of Lee Kuan Yew: Lessons for aspiring countries. https://www.rsis.edu.sg/rsis-publication/cms/co15073-economic-legacy-of-lee-kuan-yew-lessons-for-aspiring-countries/.

Restrepo, J. D. C., & Morales-Pinzón, T. (2018). Urban metabolism and sustainability: Precedents, genesis and research perspectives. *Resources, Conservation and Recycling*, *131*, 216–224.

Rocha, Z. L. (2011). Multiplicity within singularity: Racial categorization and recognizing "mixed race" in Singapore. *Journal of Current Southeast Asian Affairs*, *30*(3), 95–131.

Rut, M., & Davies, A. R. (2018). Transitioning without confrontation? Shared food growing niches and sustainable food transitions in Singapore. *Geoforum*, *96*, 278–288. doi: 10.1016/j.geoforum.2018.07.016

Shatkin, G. (2014). Reinterpreting the meaning of the 'Singapore model': State capitalism and urban planning. *International Journal of Urban and Regional Research*, *38*(1), 116–137. doi:10.1111/1468-2427.12095

Shen, Y., Zhang, X., & Zhao, J. (2017). Understanding the usage of dockless bike sharing in Singapore. *International Journal of Sustainable Transportation*, *12*(9), 686–700.

Sigdyal, A. (2018). A transformational leader: Lee Kuan Yew. https://www.scribd.com/document/234300722/Transformational-Leader-Lee-Kuan-Yew. Accessed on June 26, 2019.

Silva, B. N., Khan, M., & Han, K. (2018). Towards sustainable smart cities: A review of trends, architectures, components, and open challenges in smart cities. *Sustainable Cities and Society*, *38*, 697–713.

Singapore Business. (2018). Singapore beats Hong Kong as top Asian location for MNC headquarters. https://sbr.com.sg/markets-investing/news/singapore-beats-hong-kong-top-asian-location-mnc-headquarters. Accessed on June 25, 2019.

Singapore Land Authority (SLA). (2018, May 11). Licensing of digitised land information. https://www.sla.gov.sg/Services/Street-Map/Licensing-of-Digitised-Land-Information.

Singapore Police Force. (2016). Police news release: Annual crime brief 2016. Singapore Police Force. https://www.police.gov.sg/~/media/spf/files/statistics/20170210_annual_crime_brief_2016.pdf.

Singapore Police Force. (2017). Police news release: Singapore remains one of the safest cities in the world. Annual Crime Brief 2017, 1–13. https://www.police.gov.sg/~/media/20180203_stats_annual_crime_brief_2017_stats.pdf.

Singapore Police Force. (2018). Mid-year crime statistics for January to June 2018. *Police News Release*, 1–8. https://www.police.gov.sg/~/media/20180823_stats_mid_year_2018_crime_statistics_stats.pdf

Soper, T. (2015). Lee quoted in "The green legacy of Lee Kuan Yew". August 29. http://www.tomsoperphotography.com/blog/2015/3/29/the-green-legacy-of-lee-kuan-yew. Accessed on August 1, 2019.

Southworth, M. (2016). Learning to make liveable cities. *Journal of Urban Design*, *21*(5), 570–573.

Srinivasan, B., & Negandhi, A. (1985). Hedging in currency futures: A Singapore dollar case study. *Asia Pacific Journal of Management*, 2, 170–179.

Stuart, N. (2015). Savior king or ruthless tyrant? An Aristotelian analysis of Lee Kuan Yew's legacy. October 15. https://voegelinview.com/savior-king-or-ruthless-tyrant-an-aristotelian-analysis-of-lee-kuan-yews-legacy/. Accessed on June 29, 2019.

Tan, C. (2015). Lee Kuan Yew leaves a legacy of authoritarian pragmatism. March 23. https://www.theguardian.com/world/2015/mar/23/lee-kuan-yews-legacy-of-authoritarian-pragmatism-will-serve-singapore-well. Accessed on June 28, 2019.

Tan, K. G., Chuah, H. Y., & Luu, N. T. D. (2018). A case study on Malaysia and Singapore. *Competitiveness Review*, *28*(2), 172–193. doi: 10.1108/cr-09-2017-0062

Tan, R. B. H., & Khoo, H. H. (2006). Impact assessment of waste management options in Singapore. *Journal of the Air & Waste Management Association*, *56*(3), 244–254.

Tan, Y. S., Lee, T. J., & Tan, K. (2009). *Clean, green and blue: Singapore's journey towards environmental and water sustainability*. Singapore: Institute of Southeast Asian Studies.

Temasek. (2018). The great energy transition: Challenges and opportunities for transformation. Singapore. Retrieved from https://www.ecosperity.sg/content/dam/ecosperity/en/reports/The-Great-Energy-Transition_Challenges-and-opportunities-for-transformation.pdf. Accessed on June 27, 2019.

Teo, S. (2014). Political tool or quality experience? Urban livability and the Singaporean state's global city aspirations. *Urban Geography*, *35*(6), 916–937.

Teo, S. C. (2014). Sustainability, Singapore style: Creating a liveable city that harnesses technology to drive the economy. *Experience Singapore*, (50), April–June. Singapore.

Ter, K. L. (2018). Singapore's cybersecurity strategy. *Computer Law & Security Review*, *34*(4), 924–927.

The Economist Intelligence Unit. (2017). Safe cities Index 2017: Security in a rapidly urbanising world. *The Economist*, pp. 1–44. http://safecities.economist.com/safe-cities-index-2017.

The Economist Intelligence Unit. (2018). The global liveability Index 2018: A free overview. *The Economist*. https://

pages.eiu.com/rs/753-RIQ-438/images/The_Global_Liveability_Index_2018.pdf. Accessed on June 28, 2019.

The World Bank. (2019a). GDP per capita (Current US $). https://data.worldbank.org/indicator/NY.GDP.PCAP.CD?end=1975&locations=MY-SG-Z4&start=1960.

The World Bank. (2019b). GDP per capital (Current US$). Data. https://data.worldbank.org/indicator/NY.GDP.PCAP.CD?locations=MY-SG-Z4&start=1960&end=1965.

Ting, K. (2019). *Singapore 1819: A living legacy*. Singapore: Talisman Publishing Private Limited.

Tng, S., & Tan, S. (2011). *Designing our city: Planning for a sustainable Singapore*. Singapore: URA.

Tortajada, C., & Hongzhou, Z. (2016). Food policy in Singapore. *Reference Module in Food Science* (2017), 1–7. doi:10.1016/B978-0-08-100596-5.21083-4

Tortajada, C., & Joshi, Y. K. (2013a). Water demand management in Singapore: Involving the public. *Water Resources Management*, *12*(8), 2729–2746. doi:10.1007/s11269-013-0312-5

Tortajada, C., & Joshi, Y. K. (2013b). Water Resources management and governance as Part of an Overall Framework for growth and development. *International Journal of Water Governance*, *1*(3–4), 285–306.

Tortajada, C., & Joshi, Y. K. (2014). Water quality management in Singapore: The role of institutions, laws and regulations. *Hydrological Sciences Journal*, *59*(4), 1763–1774. doi:10.1080/02626667.2014.942664

Trading Economics. (2019). Home ownership rate. https://tradingeconomics.com/country-list/home-ownership-rate. Accessed on June 27, 2019.

Trindade, A., & Costa, M. (2011). Police brutality in Brazil: Authoritarian legacy or institutional weakness? *Latin American Perspectives*, *38*(5), 19–32.

Tullock, G. (1987). *Autocracy*. Dordrecht: Martinus Nijhoff Publishers.

Turnbull, C. M. (2009). *A history of modern Singapore, 1819–2005*. Singapore: NUS Press.

Ubaldi, B. (2013). *Open government data: Towards empirical analysis of open government data initiatives*. Paris: OECD.

UN Environment. (2016, October). Our planet: The Singapore story. http://web.unep.org/ourplanet/october-2016/articles/singapore-story.

UN Environment. (2018). *Sustainable urban infrastructure transitions in the ASEAN region: A resource prespective*. Nairobi: United Nations Environment Programee.

UNCTAD. (2018). *World investment report 2018- investment and new industrial policies: Country fact sheet*. Singapore: United Nations Conference on Trade and Development. https://unctad.org/sections/dite_dir/docs/wir2018/wir18_fs_sg_en.pdf.

United Nations. (2016). *The world's cities in 2016*. New York, NY. http://www.un.org/en/development/desa/population/publications/pdf/urbanization/the_worlds_cities_in_2016_data_booklet.pdf. Accessed on October 29, 2019.

URA. (2006). *Urban redevelopment authority annual report 06/07*. Singapore. https://www.ura.gov.sg/-/media/Corporate/

Resources/Publications/Annual-Reports/PDFs/Annual Report_2006-2007.pdf. Accessed on October 30, 2018.

Verweij, M., & Pelizzo, R. (2009). Singapore: Does authoritarianism pay? *Journal of Democracy*, *20*(2), 18–32.

Wee, H.-L. (1995). Redefining Hinterland: The Singapore experience. *Middle States Geographer*, *28*, 68–73.

Wilson, E. O. (1984). *Biophilia*. Cambridge, MA: Harvard University Press.

Winter, J. (2017). Robert Mugabe: Is Zimbabwe's ex-president a hero or villain? November 21. https://www.bbc.com/news/world-africa-23431534. Accessed on June 27, 2019.

Wolman, A. (1965). The metabolism of cities. *Scientific American*, *213*(3), 179–190.

Wong, J., & Fook, L. L. (2016). *The challenge of making cities liveable in East Asia*. Singapore: World Scientific Publishing Co.

Woo, J. (2017). Tanjong Pagar terminal: An icon of the Port of Singapore. https://www.straitstimes.com/business/an-icon-of-the-port-of-singapore. Accessed on June 24, 2019.

World Bank. (2010, December). *Cities and climate change: An urgent agenda. Urban development series knowledge papers* (Vol. *63704*, No. 10). Urban development series knowledge papers. Washington, DC: World Bank.

Yeo, S. (2016, December 31). Here's how Singapore's education system has changed from the 1960s to now. https://www.youngparents.com.sg/education/heres-how-singapores-education-system-has-changed-1960s-now/?slide=5. Accessed on June 28, 2019.

Yew, L. K. (2000). *From third world to first: The Singapore story: 1965–2000*. New York, NY: Harper Collins Publishers.

Yew, L. K. (2015a). *From third world to first: The Singapore story, 1965–2000: Memoirs of Lee Kuan Yew*. Singapore: Marshall Cavendish Editions.

Yew, L. K. (2015b). *The Singapore story (Student edition): Memoirs of Lee Kuan Yew*. Singapore: Marshall Cavendish International Asia Private Limited.

Yong, T. T. (2019). Looking back at 700 years of Singapore. *Biblioasia*, *14*(4). http://www.nlb.gov.sg/biblioasia/2019/01/23/looking-back-at-700-years-of-singapore/. Accessed on June 23, 2019.

Yuen, B. (2007). Squatters no more: Singapore social housing. *Global Urban Development Magazine*, *3*(1), 1–22.

Ziegler, A. D., Chan, F. K. S., Chuah, J., Dabrowski, M., & Varis, O. (2018). Towards resilient flood risk management for Asian coastal cities: Lessons learned from Hong Kong and Singapore. *Journal of Cleaner Production*, *187*, 576–589.

INDEX

Printed in the USA
CPSIA information can be obtained
at www.ICGtesting.com
LVHW020928201023
761597LV00017B/2300

9 781839 821073